The Eloquent Engineer

Every engineer's—and technical professional's—guide to creating and delivering compelling presentations for even the most non-technical audiences.

Jun Echevarria P.Eng.

Bob Serrano

Table of Contents

Foreword v
Introduction viii

PART I   The right mindset 1
              Chapter 1: Accept your strengths and weaknesses 3
              Chapter 2: No excuses 5
              Chapter 3: Know your purpose, audience and message 11
              Chapter 4: PowerPoint is just a tool 19

PART II   Understanding understanding 23
              Chapter 5: Brain matters 25
              Chapter 6: Picture this 31

PART III Simplify, simplify, simplify! 39
              Chapter 7: Easy on the eyes 41
              Chapter 8: Easy on the mind 53
              Chapter 9: Easy charts 65
              Chapter 10: Typos, inkblots and other horrors 79

PART IV   Effective delivery 83
              Chapter 11: Presentation traps 85
              Chapter 12: Team presentations 91
              Chapter 13: Practice, practice, practice 95
              Chapter 14: Stage fright 103
              Chapter 15: Your first two minutes 107
              Chapter 16: Let's have a conversation 111
              Chapter 17: Power tools 117
              Chapter 18: Speak their language 129
              Chapter 19: Bad practices 133
              Chapter 20: Allies in the audience 139
              Chapter 21: Closing act 143

Epilogue 146
Sources 148

About the authors

Jun Echevarria P.Eng.

Jun is a professional engineer in Alberta, Canada. He has a Master of Science in Electrical Engineering and an MBA, as well as professional credentials in project management (PMP) and business analysis (PMI-PBA).

Before migrating to Canada in 2011, he worked as an electrical engineering professor and as a senior executive in various roles in industry and government. Teaching honed his skills at simplifying and being clear. As an executive, he learned to translate technical things in terms that management could understand. He draws the book's lessons, insights and stories from experiences that bridge engineering and management.

Bob Serrano

Bob is a creative adviser for a top-tier multinational project delivery company based in Sydney, Australia, with Canadian headquarters in Calgary, Alberta where he's worked for the past 10 years in the business development group. He also worked as a creative director for a Calgary-based marketing design company while simultaneously running his own freelance design/production outfit, bobserranocreative.com.

Prior to moving to Canada in 2004, Bob ran his own boutique audiovisual production company. He also worked as an art director for a high-end lifestyle magazine and taught corporate identity and logo design, computer-based illustration, and non-linear video editing while writing original music for radio and TV ads.

Foreword

This book slowly formed in my mind for five years before I actually started writing it. I've had years of presentation experience as an engineering professor, and in various executive roles. I felt I had lessons to teach, insights to share, stories to tell. When I finished my first complete draft, I asked a few friends to read it.

One draft went to my co-author, Bob. Bob is a creative adviser in a large global engineering company. For the past ten years, he's been creating presentations for engineers— translating technical proposals into compelling business presentations.

We're friends, Bob and I. We've known each other since we were teenagers in Manila singing in a choir. Funny that we both ended up living in the same city in Canada.

Bob had a lot to say about my draft. Turns out he passionately agreed with many of my points, and had his own views and stories to add about engineers doing presentations.

For two hours and over beer, we had a lively discussion at my home. All throughout, I found my mind playing with the thought that Bob's insights, enthusiasm and skills could raise the book from being a collection of my insights and stories, to something that I hope could become a lasting reference for engineering students and professionals alike.

He offered to do the book cover for me. I countered with a proposal—would he consider to be my co-author?

This book is a collaboration. While much of it is written from my original perspective—that of an engineer speaking to fellow engineers—you are actually benefiting from a fusion of our combined thoughts and experiences.

Jun Echevarria

When Jun approached me about this project, I'll admit I was hesitant. What do I know about writing a book? Turns out, this was Jun's first attempt at penning one himself so that assuaged some of my fears.

The more I got into the material, the more it resonated with me. I help prepare, on average, about two presentations a week, mostly for engineers—sometimes for executives in leadership roles. My background in mainstream advertising and visual communications has taught me enough to see the challenges that lay ahead upon reading first drafts that were plunked on my desk. I remember thinking to myself in the past how I could write a book about all the 'stuff' I had to do to whip these presentations into shape and now, here we are. Serendipity rocks!

Jun and I have fused our personas into one to simplify matters for the reader. It's written in the first person so think of it as sage advice from someone with close to 60 years of practical, combined experience in teaching, authoring, designing and delivering presentations. Most of the stories are his, but I did have a couple here and there to add.

I personally think our combined skills and experiences make for an approachable resource for engineers, from the greenest to the most experienced.

Enjoy.

Bob Serrano

" **Eloquence** is an art of saying things in such a way that those to whom we speak may listen to them without pain and with pleasure, that they feel themselves interested, so that self-love leads them more willingly to reflection upon it."

Blaise Pascal

Introduction

Let's get two questions answered straightaway. Why should an engineer be able to deliver good presentations? What can you expect from this book?

Why bother?

We're engineers. Why do we need to be good at anything other than engineering? Why bother learning how to deliver great presentations?

Science and technology is our wheelhouse—mathematics, our tool of choice. We're creative, logical, trained to be methodical in problem solving. We like accuracy and precision. We're comfortable expressing abstract concepts in equations and charts. We design, invent and build things. This is who we are. This is how we think, work and talk with fellow engineers.

But unless you're among the handful of eccentric geniuses of old who could afford to be socially challenged while still able to change the world, most of us need more than just technical competence to get things done.

Very early in my career, my boss told me, "You're an excellent engineer. Too bad you know nothing about business."

Talk about sobering moments! I soon realized that as an engineer, I also needed to manage projects, understand enough about finance, collaborate and explain my work to non-engineers, and (gulp) write a proposal.

The reality is, there are a whole lot of things outside of engineering—complementary skills we need to be good at—if we want to get things done.

This made me decide to become more of a generalist rather than a specialist, opting to pursue an MBA rather than a Ph.D.

Some real-world, non-engineering skills will come easier than others. Our facility with numbers makes management accounting and evaluating returns on investment relatively painless. Our methods and propensity for logic make managing projects almost second nature.

But there are other skills that may prove to be more challenging because much of the struggle lies in the very nature of being an engineer—what we know, how we've been trained to think, and perhaps, even our quirks. Many times, we forget that we know certain things that are way over most people's heads. It surprises us that some people can't grasp 'simple' things while forgetting that it took us years to master them.

There's a name for this cognitive bias—the *curse of knowledge*, where a person forgets the difficulty he had in learning a skill, and, in turn, discounts how a lesser trained person feels when he deals with one.

This curse is what we unwittingly suffer from when presenting our ideas to a non-technical audience and we find ourselves fumbling for a way to bridge the gap on-the-spot—often with awkward results.

Imagine you're presenting a technical proposal to a group of managers, all non-engineers. They get to decide whether or not to support your project and approve your budget. How would you explain something extremely technical to a group of people with no engineering background, are unfamiliar with your terminologies, but who hold the power when it comes to making the final decision?

There's really only one way.

You need to present your proposal in a language that *your audience* understands. It's known as the principle of attunement. By using words that are familiar to *them*, you can better guide *their* thought process to align with yours—make it easy for *them* to follow where you're going and hopefully lead *them* to the same conclusions.

Are you seeing a pattern here?

When it comes to making great presentations, it's all about *them*. In other words, it's YOU who needs to adjust.

This book will help you understand and teach you how to make those adjustments. It's written by an engineer for engineers. Because we share similar backgrounds and likely have encountered similar circumstances at work, I'm hoping this commonality will make the learning easier. If you think about it, I'll be using the very same principle of attunement with you. I will attempt to teach you things you may not have heard of yet, but by using terms and examples *you* understand, I have better odds of getting through to you! Funny how that works.

What can you expect from this book?

Let's talk about what you should and shouldn't expect from this book.

First, let me clarify what I mean by "presentations." I'm not talking about the kind you prepare for social occasions like your daughter's wedding or motivational speeches at sales rallies. I'm talking about formal, prepared business presentations intended for a client, your colleagues, your boss, a management committee or board of directors. These are presentations designed to inform, explain or persuade.

Second, this book won't turn you into an entertainer, a comedian or a charismatic, motivational speaker. Rather, the book will help you achieve clarity of message and intent. It will help you become more understandable, particularly by an audience of non-engineers. It will help you deliver your message with more punch. You will be better prepared which will, in turn, give you more confidence.

Third, this is not a book about Microsoft PowerPoint or Apple Keynote, albeit, many of the lessons assume that you will be using presentation software of some kind and that you will be designing slides as visual aids. You will recognize and use presentation software as a tool—a means to an end—not the center of your presentation.

Lastly, don't expect to become a master presenter even after reading the book from cover to cover. Each lesson is designed to incrementally improve your skill, but only after you've understood and practiced the principles behind each chapter.

This book is driven by stories, mostly based on real events. The power of storytelling is one of the lessons in this book, but that's getting ahead of myself.

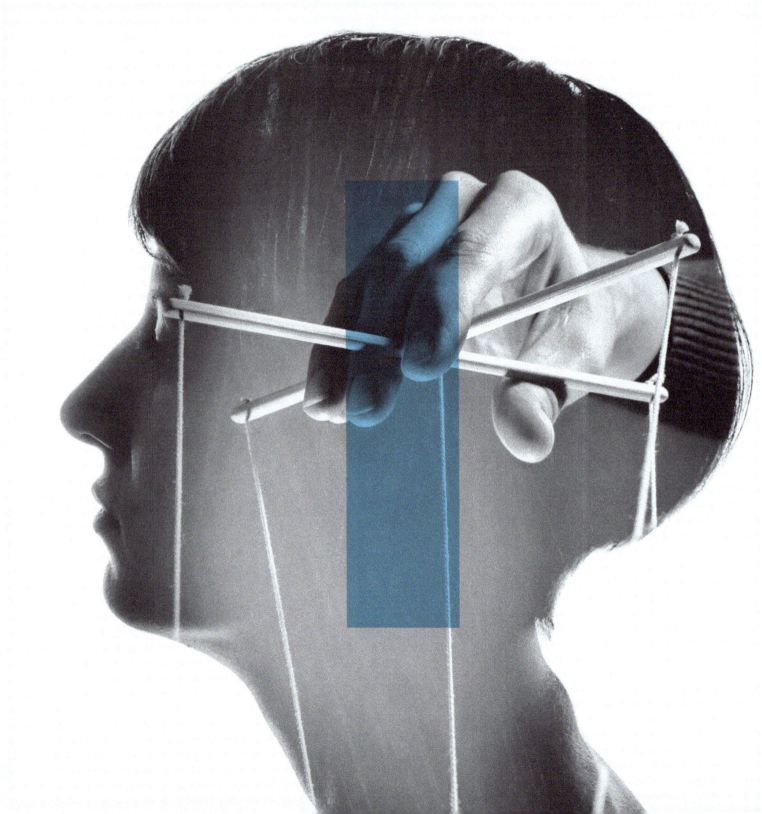

PART I

The right mindset

Taking stock of your strengths and weaknesses, while fully understanding what you can and cannot control, will better equip you for what lies ahead.

CHAPTER 1

Accept your strengths and weaknesses

A typical engineer will have built-in strengths and weaknesses when it comes to presenting.

Let's start with our strengths. Technical presentations will typically have a strong analytical component that may include technology descriptions, designs, data, analysis of options and recommendations which are arrived at with rigor and logic. Engineers generally do well in this department, having been trained in the engineering method of solving problems, and the use of science, technology and mathematics. If the only people you had to present to are your fellow engineers, and you had no time constraints, then you're probably good to go.

On the other hand, if you had to present to an audience with no engineering background whatsoever, then your weaknesses could be painfully exposed.

Here are a few of the challenges.

Because your audience doesn't share your engineering background, they wouldn't be familiar with terminologies and acronyms that you use every day—it's all jargon to them. Imagine channel surfing on TV and accidentally landing on a foreign language channel. What would you do? You'd likely switch channels, or, in the case of a presentation, tune out.

Remember how we engineers love accuracy and precision? Another potential weakness. Why? Because this often translates into presentations that have way more detail than most audiences actually need—or have the time, patience and interest to digest.

Often, presentations need to go beyond just getting an audience to understand the data and logic. There are times when you'll need to convince them that what's important to you is also important to them. You want them to be sympathetic with your messages, and comfortable enough to agree with your conclusions and give you your win.

This *emotional component* of a presentation is what will connect you with your audience and get them engaged enough to listen and agree with you. Unfortunately, engineering school doesn't prepare us to manage and leverage this emotional component.

This book assumes that you have the engineer's presentation strengths. I won't spend time teaching you to identify and evaluate options, what level of rigor is sufficient or how to approach and solve a problem. Instead, I'd like to focus on addressing our presentation weaknesses.

CHAPTER 2

No excuses

You can avoid many of the pitfalls of a bad presentation when you decide to take full responsibility for the success of your presentation. No excuses. None.

Don't blame an audience for not showing interest, being bored or unable to understand you. Don't blame the room for not having the right lighting. Traffic? Can't find parking? All preventable with enough foresight and planning.

This audience sucks!

Have you ever showed up for a talk all bright and bushy tailed, ready to conquer the world, only to have your audience start stifling yawns within the first fifteen minutes of presenting?

You might think your audience is being rude for falling asleep long before you finish, but have you considered the possibility that maybe—just maybe—you're to blame? Spending hours, even days thoughtfully preparing your speaking notes and slides doesn't automatically guarantee an audience's interest and attentiveness.

The burden of preparation rests squarely on you, the presenter. Your audience won't prepare to listen for the simple reason that they don't have to. Most times, the extent of their responsibility is showing up on time. Besides maybe knowing the title of your presentation ahead of time, chances are, they don't have a clue what to expect.

You prepare for a presentation because you want something to happen. Maybe you need to inform or explain. Oftentimes, presentations are made to persuade the right people to support your point of view, approve a proposal, or just buy something. If you don't get what you want from your audience, then you've failed. You, not your audience.

Let me start with the first of many stories.

I was assigned oversight of the IT Department of an organization. The head of the department talked about a struggle he was having with the board.

> " *Ten years ago, our company's IT system was the best there was. Every year since, I've presented an IT strategy and a proposal to the board designed to keep our system running at peak performance.*"

> " *And how did that go?*" I asked.

> " *Not good. These past few years, I've been unable to get approval for the capital budget I need to implement upgrades. The machines are now old. Maintenance parts are getting harder to source. Processing performance has slowed down. Unscheduled maintenance has been getting more frequent.*"

" What usually happens when you present your proposal?"

" I go through my slides and then get asked all these questions. I answer all of them but in the end, only the operating budget gets approved. Much of the capital budget keeps getting deferred."

" Why is that?"

" I don't think the board really understands the strategy I've presented. I believe they think the capital proposed isn't necessary."

" Again, why do you think that is?"

" They just don't understand IT. That's what I think."

He continues, "Now, I feel I'm being held responsible for the sorry state of our IT system. It's not my fault they don't understand IT. The IT strategy and plans are well thought out and defensible."

This failure to communicate technical things to a non-technical audience is common among engineers and technology professionals. There's an obvious gap in communicating how we think and what needs to be prioritized. He was clearly frustrated and felt he had done all he could.

" Don't the directors have a responsibility to make an effort to try and understand my presentation? I mean, I shouldn't be the only one responsible for this, right?"

Umm … actually, you kind of are. The one presenting is responsible for ensuring that the audience understands the message. Accepting this responsibility is an important first step. You don't get to blame the audience.

Does your audience have any responsibilities?

They gave you their time and the chance to make your pitch. They will likely be attentive for the first few minutes. After that, you alone are responsible for keeping them interested, getting them to fully understand the issues, and making things happen.

Is there a benefit to taking this uncompromising, harsh position? How does this "take no prisoners" approach help you make better presentations?

For one, it will cause you to contemplate extraordinary actions you wouldn't have otherwise considered had the responsibility for success been shared. It will push you to dig deeper into the reasons for the board's unreceptiveness. Maybe ask one of the directors what went wrong. Make a supreme effort to translate complex, technical concepts into terms that are familiar and easy to understand. You might find yourself consciously dropping jargon in exchange for plain speak and real-world metaphors. Anything to keep them tuned-in and engaged.

It's a fundamental change in perspective, for sure—a huge but meaningful 'rethinking' that, I guarantee, will result in more successful presentations.

You can't blame Murphy's Law either

Murphy's Law: "If anything can go wrong, it will." An extended version adds "…at the worst possible place, in the worst possible way and at the worst possible time."

Many good presenters dismiss Murphy, until such time that he makes a very personal, unexpected appearance. As one who's learned his lessons (plural!), might I suggest five ways to avoid getting caught. If your job depends on it, listen up.

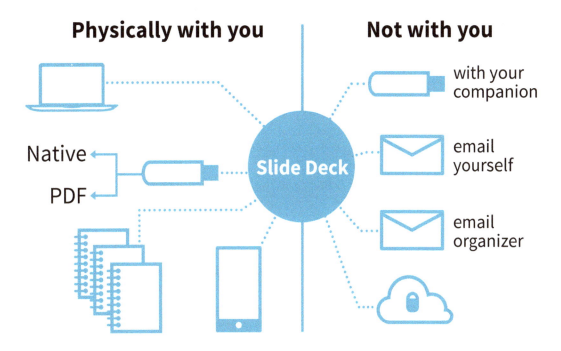

Physically with you

Native
PDF

Slide Deck

Not with you

with your companion

email yourself

email organizer

1. **If you're using a slide deck, bring multiple copies of the final version to the venue.** Each copy should come in two flavors—a native version and a flattened PDF version. In case the native version doesn't work, use the PDF—you lose your fancy transitions, but at least your fonts are automatically bundled in. Have copies on your laptop, another on a flash drive. If you have a companion, give him another flash drive copy. Send a copy to the organizer. Attach copies to an email and send it to yourself. Have copies on your smartphone. Prepare a few printed sets for good measure. If you have secure cloud storage, use it.

2. **Know what computer make and operating system will be used at the venue ahead of time**— never assume. It's one phone call or email. As much as possible, use standard fonts that are pre-installed in most computers. As of this writing, Macs and Windows operating systems combined make for 97%

of all installed computers in the world. Some popular fonts that can be found in both are: Franklin Gothic, Palatino, Tahoma, Verdana, MS Sans, Arial, Trebuchet, Verdana, Georgia, Times New Roman. The last five on the list can be found in most Linux machines.

If you must use other fonts, going the PDF path might be your safest option. Just be sure you export to an early version.

A company was trying out a mixed policy for Office software licenses; Microsoft Office for managers and up while regular staff used OpenOffice, an open-source office suite.

I had a slide presentation sent in advance for a board meeting. A board staff loaded my file on his laptop which he intended to use for all the presentations that day. I arrived just in time for the meeting.

When my presentation started, texts and images for many slides didn't display as I had intended. Font sizes were erratic. Text would, at times, overlap with each other or with an image. It wasn't my best performance.

I later learned that the staff had OpenOffice on his laptop which didn't have installed some fonts that I had used. I got fancy with my font choices and OpenOffice automatically replaced my fonts with what it had. OpenOffice might also have operated a bit differently from Microsoft Office.

Deadly mismatches

If you're required to use a shared system (this is very common in tradeshow scenarios), it's not enough to know what computer platform is being used. Find out what version of operating system is in place. Say, it's Windows 7 using Microsoft Office. Find out what *version* of Office. This is particularly important if you have embedded videos. As of this writing, MP4 has become the popular standard. It just so happens, anything earlier than Office 2013 can't play them. If you've authored your deck in an older system that uses WMV, you may need to convert all your videos to another format or risk getting stuck on stage not knowing what to do when your embedded videos refuse to play. Details, details, details—pay attention to them.

3. **Bring your own fully charged laptop** and don't forget the power adapter and video cable. For laptops using proprietary, esoteric or non-mainstream audiovisual sockets, be sure to bring the special dongles required to convert them to more standard connectors. Most modern projectors will take VGA or HDMI inputs so make sure you have the right cables and dongles. Be ready to switch over to your laptop at a moment's notice. If your deck has embedded video with sound, make sure you have the right audio cables to connect to the venue's sound system. Good professional audiovisual rental companies are equipped to handle multiple audio jack formats—but don't assume. One phone call is all it takes to make sure. While HDMI cables are supposed to carry audio signals, some setups only use it for video, requiring a separate audio connection. These are important details you can't afford to ignore.

4. **Arrive early at the venue**—really early. Arriving early gives you more breathing room—a chance to check the computer setup, the sound system, the size of the venue.

Is there a working computer to run the presentation? Is there a working screen and projector? Will there be a remote clicker? A second, presenter-facing monitor? Is a sound system available and ready? Do you have a Plan B in case you run into something unexpected? Don't depend on the designated audiovisual supplier to provide the necessary cables and dongles. Always bring your own.

Arriving early allows you to test-run your presentation before the audience arrives. Check that the correct version of your presentation is loaded on the actual computer you'll be using. Do a quick test run if possible. Click through all the slides. If there are issues that can't be corrected immediately, insist on running your presentation from your laptop.

5. **Watch out for Murphy's tricks**—there are many. Murphy knows where your weaknesses are and will try to trip you up.

   My staff organized a meeting off-site and brought a laptop, screen and projector. The audience was already in the room while my staff set up. My staff then realized that the electrical outlet was far from the projector and the power cable wasn't long enough. While they scrambled to find somebody in the facility that could help, I pulled out an extension cord from my bag. You could see the relief (and surprise!) on my staff's faces.

   Why did I have one in my bag? We were setting up in an unfamiliar place. I had an uneasy feeling Murphy would be waiting. And I was right!

Murphy still catches me off-guard from time to time—but not too often.

You might think I give Murphy way too much credit. Maybe so. But it's all in keeping with taking full responsibility for the success—or failure—of every presentation I deliver. It's also about staying two steps ahead, especially when it comes to presentation hiccups that are thoroughly preventable.

D-day checklist

Venue address:

_____

Function room/ballroom name:

_____

Point person/cell number:

_____

Call time/ingress: (arrive an hour earlier)

_____

Slide deck copies/backups

with you:	not with you:
☐ memory stick	☐ memory stick w/ colleague
☐ native file	☐ email to yourself
☐ PDF	☐ email to the organizers
☐ laptop	☐ cloud (e.g., OneDrive, Dropbox)
☐ printed copies	
☐ smartphone	

Peripherals/cables/dongles:

☐ power brick   ☐ power bar/extension cord

video cables:

☐ VGA   ☐ DVI   ☐ HDMI

☐ MicroDV ☐ USB-C   ☐ Composite video

☐ clicker/laser pointer   ☐ Ethernet cable

☐ Wifi/password: _____

If presenters are sharing a computer:

Platform: _____

Operating system: _____

Version: _____

Presentation software: _____

Version: _____

CHAPTER 3

Know your purpose, audience and message

Every presentation has a purpose, an audience and a message. This message, when effectively delivered to your audience, is what leads you to achieve your purpose. If you want to be effective, you must be clear on all three.

What is your purpose?

The presentation's purpose is the reason a group of people are coming together and setting aside time to listen to you.

Do you know why you are presenting? In every presentation, your audience has expectations and you have an objective. Meeting the audience's expectations and meeting your objective are the reasons for making a presentation.

Sometimes, a presenter dives into a presentation without clarity on the reason for his presentation. I was guilty of this once.

In the mid-90's, interest in the Internet was growing. More and more people wanted to get connected. Back then, getting connected wasn't as easy as it is today. You had to buy an external modem. You connect the modem to your computer's serial port and to a phone line. You then run a program that connects your computer via the modem and the phone lines to an Internet Service Provider or ISP. For many, the thought of connecting a modem (a what?) was daunting. And how does one even begin selecting an ISP?

A company decided to offer afternoon seminars about the Internet. The seminar's newspaper ad included a bullet that said "Learn how to get connected".

I was the speaker. I had a lot to say about the Internet— how it started and available services like email, FTP and early browsers. I talked about online etiquette (a big thing back then) and where the technology was going. All very interesting, to me at least.

When I was wrapping up, a young lady raised her hand.

" But…" she stammered, "How do I get connected?"

I was awash with embarrassment. I had assumed that what was interesting to me would be interesting to the audience. It was just then that I realized that many in the audience paid good money and came to "learn how to get connected"—as advertised.

They were hoping to get specific questions answered, like what equipment was needed, where to buy it and how much it would cost. How would they go about selecting an ISP, which were the good ones and if the ISP could help with setting up.

Their expectations were pretty basic but I didn't take the time to ask and be clear about what's important to them. What was I thinking talking about Internet etiquette? My presentation completely missed the mark—by a mile. You have to be very clear on why you're presenting. If you can't picture or define what a win looks like at the end of a presentation, how can you design a presentation that leads to that win?

Usually, a win is defined two ways—a win for you and a win for the audience. You have a purpose for doing the presentation. But there has to be something in it for the audience as well, especially if they paid to be there. What do they get out of it?

Describe the win. Everything about the presentation—messages, slides, delivery style—should lead you and the audience towards that win. You close your presentation with the win in mind.

If your objective is simply to provide an update on a project, end your presentation by asking your audience if they have questions on the status.

If your objective is to get approval for a proposal, end your presentation by asking for approval of the proposal. "I would like to request for approval of this proposal." Then wait for an answer.

If you don't ask, you won't get an answer.

Always end with your purpose in mind. Close with your win.

Really know your audience

We've established earlier that you will need to define a presentation win not just from your point of view, but from the point of view of your audience. Your audience will have its own reason for being at the presentation. You need to know and address that reason.

That said, when crafting your message and planning your delivery, you have to know much more about your audience—other than the reason they're there.

Earlier, we hinted at what this book's target audience might be. Since we're focusing on formal, prepared business presentations, we expect our audience to be "potential or existing clients, colleagues, bosses, a management committee or board of directors." We're also keen on being able to effectively present before a non-technical audience.

So, exactly what else do we need to know?

A seasoned consultant recalled how, as a young and eager consultant, she was asked to deliver an hour-long presentation for corporate secretaries.

As she prepared, she had in her mind the image of young secretaries, mostly women who provided administrative support for their managers.

Imagine her horror when, upon entering the room, she was met by an audience of mostly middle-aged men in suits. These men were executives appointed to various boards as Corporate Secretary.

Most times, we need to be more specific. The more we know our audience, the better we can tweak our purpose, craft the message and plan the delivery. We'll need to ask a few more questions.

My first few questions would be, how big is the audience? Am I presenting before an intimate group of just two or three people? Is it a boardroom presentation for maybe ten people? Or is it more of a conference presentation to a group of 200 people?

If it's a small group with known attendees, I would ask about their current roles in the organization. Am I presenting before decision-makers? Influencers?

When you get to present before decision-makers, especially senior executives or a board, be prepared to answer questions, be clear, succinct and make your case quickly and ask for a decision. You might not get another window to get before this group. With this type of audience, it's usually a good idea to keep overly detailed and granular slides hidden, but be ready to bring them up if, and only if, they ask about those details. There's a good chance that they won't.

If your purpose is to ask for approval, and your listeners are not the decision-makers, it would be futile to ask for approval, wouldn't it? If they are in a position to make recommendations or influence decision-makers, then what you're asking for is support, not approval. Is there anything else I need to know about this group before I make a pitch?

I was involved with an engineering team developing an energy management plan for the corporation. Among the strategies we considered was the use of energy performance contracts (EPCs).

EPCs allow corporations to undertake energy efficiency projects without the need for up-front capital by partnering with a private company. The private company provides the capital and much of the needed engineering expertise. The EPC would generate energy cost savings to pay for the project and then some, benefiting both entities.

A long-time energy manager expressed concern about putting forward an EPC recommendation to a group of senior managers, citing how a past EPC had failed and the company got burned. EPC had become a bad word since then.

The team argued that we had a solid business case. Besides, that unfortunate EPC happened a long time ago.

The manager seemed unconvinced. "Half of the senior managers we will be presenting to signed off on that failed contract."

True enough, the EPC part of the presentation gained zero traction with the senior managers.

Influencers

Occasionally, a small group will have an alpha, a key member that can influence the thinking of the entire group. That person may be a board chair, or may represent the majority stockholder. Knowing who that person is and how he or she thinks can help you tailor your presentation to get support and swing the group.

put yourself in their shoes. What are they looking for? What do you need to demonstrate clearly to get them to agree with you?

What kind of questions will be going through their minds?

Is the proposal clear, logical and well thought out? Is it thorough and accurate? Is it aligned with corporate goals and strategies? Were key stakeholders engaged? Are the commercial numbers sound? Are all risks identified, assessed and mitigable? Does the nominated team have the requisite skill and experience to ensure success? Who's the executive sponsor?

Getting inside your audience's head will help you craft a message that will resonate with them.

Large group presentations are more likely intended to inform and explain. In this situation, I would try to find out if there's anything common amongst the group, anything that can help me better connect to the group. Do they come from the same industry? Same company? Same department? Same profession? How about group demographics? Or is it a general mix?

After you get to know more about your audience, you might want to revisit your purpose and see if you should tweak or change it.

Now that we're clear on our presentation purpose and know our audience, we can work on proper messaging.

What is your message?

Every presentation has a core message. It's that piece of information, idea, emotion or conclusion that you want your audience to receive, understand, feel or agree with.

Your goal has to be that once your audience receives your message the way you intended it, you will have achieved your purpose.

There are two common mistakes presenters make when crafting the message. Either the message lacks clarity or there are just too many messages so that the main message gets lost. Your audience will either be unclear in their understanding or end up confused.

Remember, you have to work within a time limit. You should only deliver as many messages as necessary, to achieve your purpose. No more, no less.

A presentation will often have one core message, supported by a few secondary messages. Identifying this core message should be the starting point in your preparation process. Being able to clearly and concisely verbalize this core message is critical. Once you've identified your core message, you can then develop your presentation around it.

What if you have two equally important core messages? I suggest that you consider having two separate presentations, akin to being in two agenda items. Another way is to partition your talk into two distinct parts, with each part having its own core message.

I often challenge myself to try and state the core message in just a sentence or two.

Next time you start developing your presentation, start with four questions.

1. What is the **purpose** for delivering this presentation?
2. Who is my **audience** and what do I need to know about them?
3. What is my **message** to this audience?
4. How will that **message** lead to **achieving the purpose**?

The elevator pitch

How, then, do you identify and get clarity on the core message?

Some people like to brainstorm. They generate as many ideas as they can, sift through the list and start vetting until they're left with the key messages. That works for many people.

I prefer doing it a different way—granted that it may involve some amount of brainstorming, research or study. My preferred technique is crafting an "elevator pitch." It's a prepared and practiced spiel that summarizes the core of your presentation in 30 seconds—the time of an average elevator ride.

Here's the scenario. You have a proposal you've been itching to talk about with a key decision maker for some time now, but he's always on the move and hard to pin down. You get into an elevator and there he is, alone! And he's not even talking with his wife on his phone or updating his Facebook page. Now's your chance!

Good thing you're ready. You introduce yourself, deliver your spiel as if it were just a casual conversation—in under 30 seconds, before he arrives at his floor.

As the elevator doors open, he says, "That's really interesting. Call my assistant. I'd like to hear more about it."

If that's the response you get in this albeit hypothetical situation, then chances are, you've successfully crafted your core message. The whole point of the elevator pitch is to force you to concentrate on what's truly important and drop what's trivial. If you can't get it down to 30 seconds, it just means you haven't worked hard enough at it. It means the core isn't quite clear enough in your head yet.

After crafting your 30-second core message, ask yourself "What if I have two minutes, what messages can I add to better explain the core?" You then add messages to support and expand on your core.

After you have your 2-minute spiel, you can ask yourself, "Now, what if I have five minutes? Ten minutes?" The process of adding messages and slides incrementally is a great tool for creating the structure of your presentation. It will also come in handy if you suddenly find yourself with less time.

Say, for instance, you were initially told you had thirty minutes for your presentation but the schmuck presenting ahead of you took twenty minutes longer than his allotted time. The chairman approaches you and says "It's almost lunchtime. Can you shorten your presentation please? You have ten minutes."

Instead of wasting time and energy planning the untimely demise of the guy ahead of you, pull out your 10-minute pitch, and you're good to go. Don't let the sudden turn of events faze you. That's life—deal with it and move on.

To script or not to script

So, you've gone through the elevator pitch exercise. You now know your core message as well as the supporting messages. Now it's time to start constructing your presentation. What message comes first? What comes next? How do you build up your business case towards a win?

An outline is always a good place to start. Lead with the juiciest, most attractive point for your audience. It will pique interest early and show that you're not wasting their time. Then lay out the rest of the story, playing around with the sequence, until you slowly see your core message come to life, with clarity and persuasiveness.

And then what?

For some presentations (and some presenters), a detailed outline is all they need to speak on a topic or start creating slides. When pressed for time preparing a non-critical presentation, I often find myself simply breaking up my outline into slides, expanding on each point while carefully keeping time.

So, should you write a script? Let's look at some pros and cons.

Pros: Creating a script has value.

1. You get to choose the words carefully, ahead of time and say exactly what you mean with no unnecessary information. You won't have to conjure up the words as you go along.

2. A studied script minimizes the risk of rambling, because you've already thought out what you want to say.

Cons: Creating a script is not for everybody.

1. It could get tedious to write, especially for presentations that run for more than an hour. At an average rate of 150 words per minute, an hour-long presentation means you'll be writing a 9,000-word script. That's 18 pages of single-spaced type!

2. It could become a crutch while presenting, tempting you to look down at your notes too often, losing eye contact, or just reading off the sheet from time to time, since it's right there in front of you.

3. Your writing style may tend to be more formal and less conversational than your speaking style. You may come across as unnatural.

So again, should you write a script?

For special situations—short but critical presentations where you can't afford to miss any details and where timing is critical, then writing a full-on script is the best way to go. For longer presentations, I would at least script out the introduction, key insights, key messages and my closing.

If you tend to ramble, writing a script would also help. Other than these, scripting is a preparation tool, a personal strategy, a choice on what works best for you.

If you do decide to script, bring yourself—through practice—to the point where you're comfortable enough to drop your script. Read your script aloud—alone at first to see how certain words roll out more comfortably than others. With a practice partner, get feedback regarding volume, pitch, pace, intensity, gestures, facial expressions, as well as when to pause for good effect. Do this repeatedly, trimming and improving each time, until you 'own' it.

A personal note on scripting—some people are born writers, and some, born speakers. This is something I learned from my dad—a born speaker. His first drafts usually tended to sound too formal—and I recall him recording himself over and over, listening to his cadence and choice of words, editing himself mercilessly until it sounded 'just right.'

So, when writing, try to lean more toward a conversational style. Think: "words for hearing" instead of "words for reading."

Practice delivering the script until the words flow naturally. Go beyond just memorizing words. Internalize the messages and own them.

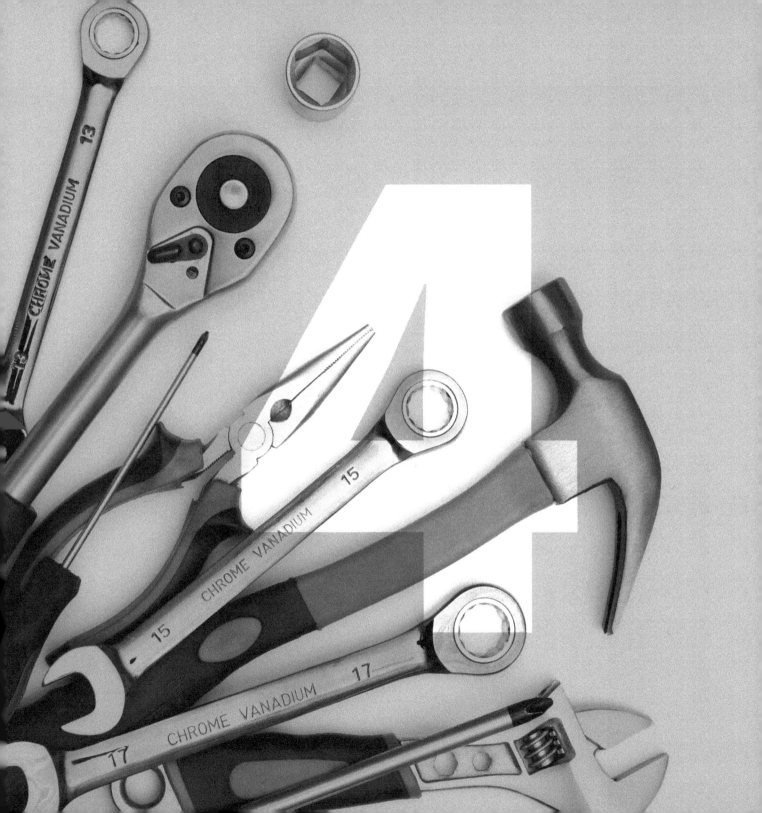

CHAPTER 4

PowerPoint is just a tool

My father was a mechanical engineer who was quite handy with tools. I grew up watching him tinker away in a workshop full of tools, assorted spare parts and junk. Later on, as an engineering student, I would work with him on father-son projects. I remember that workshop fondly as one version of heaven. It was our happy place.

He gave me two bits of advice about using tools: always use the right tool, and never blame your tools for bad work.

I'm reminded of the saying "It's a poor craftsman that blames his tools."

Very likely, you will use presentation software like PowerPoint in making presentations.

PowerPoint is a tool. Like any tool, you should know what it's good for, when to use it and when not to. When you do use it, know how to wield it properly. But never blame PowerPoint for shoddy slides or presentations. The fault lies squarely on you, the person operating the tool.

Alas, the allure of all the available bells and whistles, coupled with its ease of use have seduced many people into picking up some very bad habits that have resulted in sucky presentations. It's sad, but many presentations out there fall into this unfortunate category.

Temptation # 1:
Presentation software can make you lazy.

How often have you attended presentations where the presenter reads text off the screen—sometimes word-for-word? Can you imagine watching a new movie where the screenplay is flashed on one side of the screen, before the characters have had a chance to act out their parts? As unthinkable as that sounds, that's precisely what takes place in many boardrooms around the world. By flashing all these words on screen ahead of time, you're actually inviting the audience to tune you out and just read in advance.

For fear of forgetting what they're supposed to say next, or missing out on something important, many presenters use on-screen bullet points as a crutch. They don't do it on purpose, mind you, but because it's so widespread and seemingly acceptable, and the convenience to the presenter so obvious, people just give in to the practice. But since when was a presentation about the presenter's convenience? Didn't we already establish that when it comes to presentations, it's all about them—the audience?

Temptation #2: Creativity gone wild.

Because presentation software has placed so many 'toys' under our disposal—font styles, SmartArt, fancy transitions, animations, and the most nefarious of them all—sound effects—some people forget that they're composing a business presentation.

As amusing as it may be for your inner child, you need to be aware of the dangers of distraction.

> I remember attending a learning workshop years ago where the presenter used just about every possible transition and animation that PowerPoint had to offer, while accompanying bullet points with 'cute' sound effects as they flew in. I can barely recall the subject matter but what I do remember vividly is how annoyed I was at the end of the three days. Which is probably what made the workshop subject so forgettable—all I remember was being irritated.

Furthermore, bells and whistles can't make up for weak content. Solid content is best appreciated unadorned. It's like having white space on a page. The more white space you surround an object with, the more you pull focus onto the object.

There are lots of ways you can exercise creativity—overusing effects isn't one of them.

La semplicità è la migliore sofisticazione.

" **Simplicity is the ultimate sophistication.**"

Leonardo da Vinci

PART II

Understanding understanding

This section will help you understand how the human brain receives, processes and comprehends visible and audible messages. It helps explain the research and science behind many of the lessons in this book.

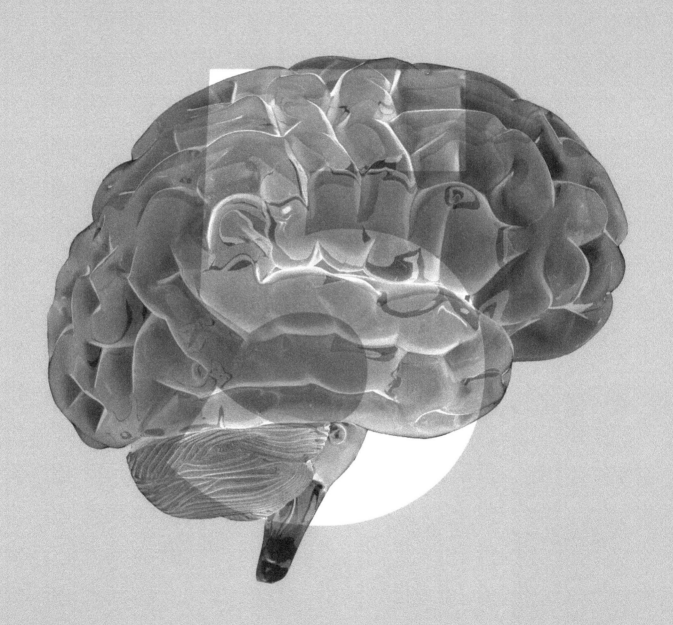

CHAPTER 5

Brain matters

Mechanical engineers would probably never make the mistake of filling the tank of a diesel engine with regular gas, or a gas engine with diesel fuel. They know the results would be disastrous.

That's because they understand how a diesel engine uses glow plugs and requires a lot more cylinder compression than a gasoline engine that uses spark plugs and has a completely different way of achieving combustion altogether. So the type of fuel we would feed each one is unique.

The same applies to the human brain. Unless we understand how this 'cranial engine' works, how it feeds on information (fuel) and how it processes that information most efficiently (combustion), then we won't know exactly how to formulate or configure the fuel it needs.

You're probably wondering why I'm devoting an entire section to understanding how the brain works in the context of communication. I mean, how complicated can it be? I've used my brain my entire life, I think I've got a pretty good handle on how it works.

Perhaps, but truth is, neuroscientists have been studying how the brain processes information far longer than most people have lived. We need to pay attention to some pretty mind-blowing facts they've discovered through the years. Facts we probably never realized and which can give us a leg up in making our presentations more effective.

While studying brain function has very little to do with engineering, the science behind the discoveries is well-founded, easy to apply in presentations and can deliver immediate, tangible results. So why not make use of it?

Speed Differentials

Imagine a weird vehicle where the wheels on one side could only spin up to a certain maximum velocity and the wheels on the other side spun a lot faster. The vehicle would obviously spin out of control with no forward direction.

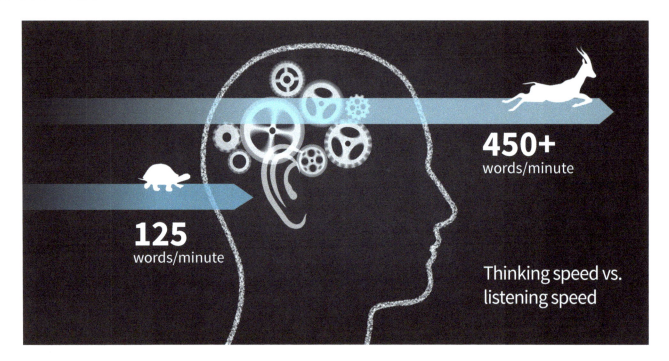

450+
words/minute

125
words/minute

Thinking speed vs.
listening speed

In much the same way, most humans are capable of listening at an average of **125 words** per minute, while the brain spins a lot faster, able to process up to **450+ words a minute**. That's a speed differential of about 75%—excess brain processing capacity "looking for something to do."

What this means is that during a presentation, while you're speaking, there's a very high potential for your audience's thoughts to wander. Considering that neuroscientists have proven that vision takes up at least 50% of our brain's resources, distractions can come in the form of extraneous elements you've unwittingly placed on slides that have nothing to do with what's coming out of your mouth. The most common examples of extraneous elements are bullet points and images you've already displayed but are not quite ready to talk to just yet.

You can't command an audience to NOT pay any attention to other stuff on the screen and just focus on that one thing you're talking to. You'd be foolish to think you could. So this leaves us with no practical recourse but to eliminate things that we're not tackling at that precise moment.

This very simple yet fundamental rule is broken or ignored in more than 90% of the average slides in a presentation. Seriously. Every time you flash more than one bullet point at a time, more than one picture at a time or quite simply, more than one message at a time, you've broken the rule and unwittingly allowed your audience's eyes and minds to drift.

I'm not making up this rule. It's all based on hard facts that scientists have proven with years of empirical data. Accept it, embrace it. Better yet, use it to your advantage.

Many presenters have this misguided notion that squeezing multiple messages into one slide somehow saves time. Fact is, the only thing you're saving is the number of slides in your deck. What you think you're gaining in time, you're actually losing big time in audience focus, understanding and retention.

Pictures vs. words

Every time audiovisual information is communicated, a kind of battle ensues in the brain of the person on the receiving end of that information. The battle is between understanding through words or through images. What gets retained? What gets forgotten?

Pictures trump words—easily.

Words and numbers are processed by our short-term memory—specifically in the prefrontal cortex, where we can only retain five to seven chunks of information at a time—five for words, six for letters, seven for digits. This is according to one of the most cited papers in psychology by George A. Miller in 1956. This is part of the reason why phone numbers only have seven digits. The same part of the brain is responsible for short-term memory. According to the same study, if you go beyond those "chunk" limits, all the extra numbers or words pretty much "go in one ear and out the other."

Pictures, on the other hand, go straight into the deep areas of the brain, controlled by the hippocampus where long-term memory resides. But not only are pictures remembered longer, they're also recognized and understood faster. In fact, some of the latest findings from MIT show that the human brain can recognize an image in as little as 13 milliseconds!

That explains why it's always easier (and faster) to just show a picture of something as opposed to describing it in words. Why do you think "show-and-tell" is so popular with kids?

Case in point: The following is a description of a square.

A square is a regular quadrilateral, which means that it has four equal sides and four equal angles (90-degree angles, or right angles). It can also be defined as a rectangle in which any two adjacent sides have equal length.

Consider the alternative:

So, knowing that pictures or images work faster and get remembered longer, how should this influence the way we author our presentations?

Simple. More images, less words. Let the words flow from your mouth, not your slides. That way, your audience will focus on listening to you, while your slides play the role they're supposed to play in the first place—visual support. In short, YOU are the presentation—not the slide deck.

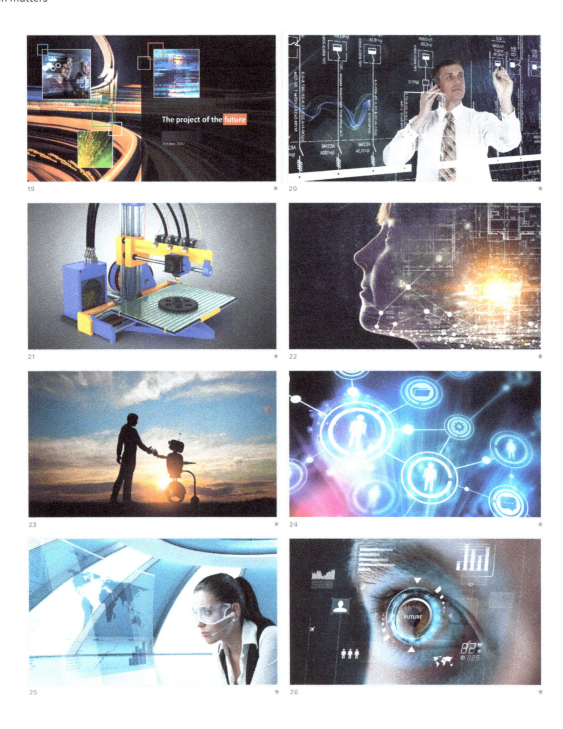

The idea of filling a deck with mostly pictures instead of words may seem daunting at first, but with practice comes confidence, and with confidence, success. A little help from an experienced graphic designer can come in handy, given their experience in not only sourcing but also editing images to suit your exact needs.

More on that later. For now, it's important to plant the seed in your head that if you use more images than words, there's a greater likelihood your audience will understand more and recall more, long after you've left the boardroom.

◄ Slide excerpts from the presentation **"The project of the future"**— an exceptional example of how a slide deck can be used effectively to enhance a talk using imagery that is both relevant and punchy, with little to no text so the audience can listen while maintaining focus on the speaker's voice and content. Used with permission from the author. Images from Adobe Stock.

CHAPTER 6

Picture this

I'm sure you've heard the saying "a picture is worth a thousand words." It's a popular expression that suggests a complex idea can be conveyed more effectively with just a single, still image compared to an exhaustive word description.

This can be a very powerful tool for presentations, but this begs a few questions for an engineer who's trying to compose a slide deck:

1. How do I choose the right image to convey exactly what I want to say?

2. Can everything really be expressed in pictures and images?

3. How about complicated technical descriptions and flowcharts that are commonplace and often necessary in presentations by engineers? Can pictures really replace those?

These are all very good questions, and the answers won't come easily. But if you're serious about making your presentations extraordinary and noteworthy, then dig your heels in and learn. There's a reason why many presentations out there aren't remembered or talked about. It's because the presenters took the easy way and just placed their words on screen, turning the presentation into a reading exercise instead of an engagement experience.

Let's tackle the first question.

How do I choose the right image?

Remember how we said in the foreword that this is not a book about PowerPoint? Well, who said that the "image" even had to be a PowerPoint slide?

A good friend of mine, a prominent water technician and process engineer, had been trying to convince people to invest in a technology he was pioneering which would improve the quality of steam in a hydrocarbons plant. This would, in turn, save operators a lot of money in maintenance cost. His past pitches were not fruitful.

He had another presentation scheduled. He asked for my help.

I asked him to deliver his presentation to me as he would his usual audience. Less than halfway through, I had to stop him—in between stifled yawns. Not only was his presentation overflowing with technical jargon from the get-go, but in his attempt to convince his potential customers of his mastery of the technology, he had jammed way too much information in the form of on-screen text. It was hard to get what his core message was.

At this point, he had an Excel worksheet up on screen, with 5 columns and about 15 rows, detailing the chemical breakdown of a facility's effluent before and after the application of his technology on a laboratory scale.

I asked him, "What are the truly problematic chemicals on this list?" He said "these three"—and pointed at them in the worksheet. Apparently, these were the three compounds that gunked up the system and cost companies millions in annual cleanup.

I then asked him, "What do these compounds actually look like in real life?"

" Oh, these three compounds combine to form a thick, shiny, black sludge that resembles tar but is water-soluble, inert and incombustible," he answered.

That gave me an idea. Forget the Excel worksheet. Get the audience to focus on that sludge!

I asked him if he had a sample, and he said he actually had a vial of it in his briefcase. Perfect! I told him, "The next time you get to this part of the presentation, take out the vial, pour some of that sticky goop onto a petri dish, put on some protective eyewear then pull out a portable blowtorch—the kind chefs use when making crème brulée. Torch the substance in front of them just to show them how incombustible it is."

He looked at me with a puzzled look. "Why? What would be the point of all that?"

" It's a powerful image that your audience would take home with them and remember. Most of them probably don't even know what the gunk actually looks like. Giving it a face, so to speak, makes the problem 'real'. Putting on the goggles and torching this substance right before them will make it more memorable— visually!"

$5M

per year in cleanup costs

" This black sludge is costing you $5 million a year in cleanup costs. My technology will completely eliminate this problem."

I added, "Conclude your presentation with a powerful takeaway message. Look them in the eye and tell them, "This black sludge is costing you $5 million a year in cleanup costs. My technology will completely eliminate this problem."

The next time I saw my friend, he flashed me a big smile, and two thumbs up.

In this story, the "picture" was the torching demonstration. If this kind of demonstration is impractical for, say, a ballroom presentation with a large audience, then do the next best thing. Have someone photograph the exercise and show it in living color, close-up, on screen. Then fill in the information by talking to the image, live.

On a simpler, more practical scale, choosing an appropriate picture shouldn't be that hard. If you're talking about the details of a project, instead of filling the screen with bullet points, try showing a picture of the actual facility instead—or a computer rendering if it hasn't been built yet. Instead of just flashing a mathematical formula on screen, show its graphical representation. Instead of lengthy descriptions of the environmental impact of an underground contaminant plume, flash a picture of the actual situation, or a graphical 3D model. If you can make the graphics interactive, that would make it even more mind-blowing.

Sometimes, in a presentation, you want to elicit a more emotional reaction from your audience, especially when speaking on broader concepts like success, safety, risk and the pursuit of excellence, to name a few. This is when you need to get a little bit more creative. If you have access to a stock photography site, it's possible to search for photos or illustrations based on genre, mood or emotion. This can be a big help when searching for just the right image.

Affective is effective

Don't be scared to engage your audience on an emotional level. Neuroscience has already proven that the memories you never forget are the ones that were accompanied by a strong emotional response at the time they occurred—whether it's sadness, elation, fear or excitement. So wouldn't it make sense that a presentation that strikes some kind of emotional chord with an audience, would likely be remembered more? Absolutely! And using a perfectly timed visual, such as a well-composed photograph, can do just that.

Think about it. What's your favorite movie? Your favorite actor? Your favorite song of all time? These are your favorites because they resonated with you on an emotional level. I'm not suggesting you get overly emotional and impassioned while presenting. Emotions come in varying degrees—but even the subtlest of emotions will register and add to overall engagement.

So, exactly what kind of emotions would you like your audience to experience during a presentation? How about curiosity?—in a process you've introduced that they've never heard of before. Confidence—in your skill as an engineer. Trust—that you're able to deliver on a promise. Relief—in knowing you hold the perfect solution to a nagging technical challenge. Best of all, excitement—at the prospect of saving a ton of money when they're convinced your proposal will achieve that very goal?

The possibilities are endless.

"Adopting this technology will take your company to new heights."

" Are you certain you're ready to take this risk?"

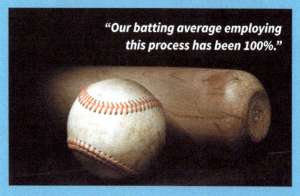

"Our batting average employing this process has been 100%."

Can everything be expressed in pictures?

If you really put your mind to it, yes—pretty much. History tells us the earliest forms of writing from Egypt and Mesopotamia were pictographic.

Now when it comes to your modern-day presentation, think about it. These pictures won't be sitting idly on the screen while your audience tries to decipher their meaning. You and your spoken words will be the center of attention—the image is mere support. How effective that support is, depends entirely on your choice of image. A great image should make your words excitingly come to life. Accompanying your words with relevant and powerful imagery will cause your audience to send those signals into the deeper recesses of their brains where long-term memory resides, instead of going in one ear and out the other.

If you don't believe that text and bullet points can be replaced with pictures, here are a few examples to get you going. We'll start with some straightforward transformations.

Specifications for Proposed Generator

- 25-370 MVA
- brushless exciters
- 3-phase electrical / air-cooled
- vibration reduction using flex plates
- IEC and IEEE/ANSI-compliant base

Instead of flashing bullet-point text describing the technical specifications of, say, a generating facility you plan to build...

Show a picture or diagram while speaking, perhaps highlighting portions as you reference them, using animation or a laser pointer. If you're worried they won't remember the numbers, assure them it will be included in a technical, leave-behind document.

$$\begin{bmatrix} r_0 \\ \theta \\ z \end{bmatrix} = \begin{bmatrix} 3+\sin t+\cos u \\ 2t \\ \sin u+2\cos t \end{bmatrix}, t=0\dots2\pi, u=0\dots2\pi$$

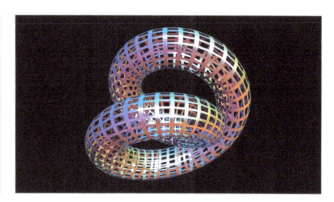

Even the brightest of engineers will be hard pressed to visualize on the fly a set of equations in multiple dimensions.

So why not do the visualization for them?

The Environmental Impact of Unsorted Waste

- Garbage mixed with plastics interferes with waste processing facilities and may also cause problems in landfill operations
- Requires large areas for disposal
- Garbage containing plastics, when burnt may cause collateral air pollution by emitting polluting gases
- Recycling industries operating in non-conforming areas are posing unhygienic problems to the environment and the surrounding communities.

Instead of going through bullets about unsorted waste...

Show an image instead and start talking about it. Your audience might not even remember much of your description but they will remember the image and the emotion it evoked. Better yet, tell a story about unsorted waste, while showing the picture.

What if I need to get technical?

No problem. As a matter of fact, the more technical a topic is, the more you need imagery to help things along.

Remember how, in the previous chapter, we had established that overloading a presentation with words and numbers will result in most of it being forgotten?

Well, highly technical presentations tend to be very wordy, often jam-packed with numbers and spreadsheets. If you want all that technical information to be understood and retained, I would suggest recouching that information in more graphical or pictorial ways to increase retention.

If you're worried they won't remember key numbers and figures, explore the option of flashing only key numbers or phrases overlayed at the precise moment (click) you are talking to it. Like so…

Spoken:

" Last year alone, (click) more than four hundred million tons of garbage ended up in landfills in the United States…"

Visual on click:

PART III

"Simplify, simplify, simplify!" Henry David Thoreau, 1854

Mr. Thoreau was definitely on to something. Always strive for ease of understanding by keeping things easy on the eyes and easy on the mind.

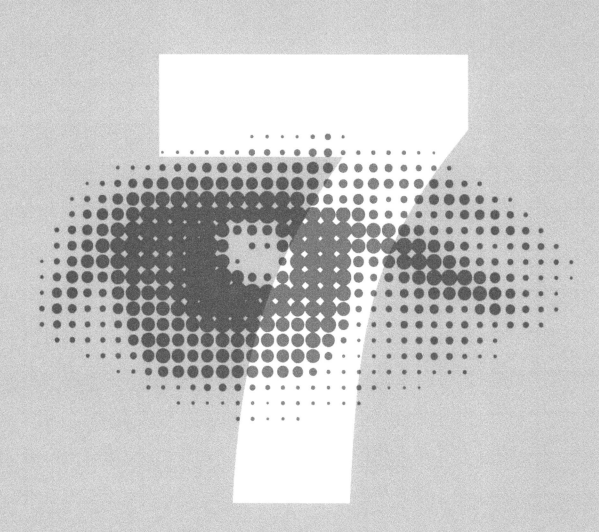

CHAPTER 7
Easy on the eyes

How many times have you watched a presentation where the presenter decides to throw a slide with about a couple of hundred words on screen? A few presenters might even introduce the slide with "you probably can't read this but …" Sound familiar?

If he were paying attention, he'd see everyone in the audience squint in unison, as if on cue. This occurrence is way more common than you might think.

The audience might not be conscious of the added effort but it's there. Every slide where your audience needs to squint adds to their collective fatigue while subtracting from their collective understanding.

I don't think presenters do this intentionally. But (and here we go again) because the practice is so widespread, people don't think there's anything wrong with it. Well, now you know, so stop it already. As a presenter, it's your job to make your audience's experience of consuming information as 'frictionless' as possible, making things easy to understand and recall.

I think we can all agree that font size and the contrast between text and its underlying background are major factors to consider.

For the figure below, up to what point can you distinguish between an 'S' and a '5' without effort?

Notice how much easier it is to distinguish between S and 5 when the font size is larger, and when there is greater contrast between the text and the background?

S5S5S5S5S5S5S5
S5S5S5S5S5S5S5S5
S5S5S5S5S5S5S5S5S5S5
S5S5S5S5S5S5S5S5S5S5S5S5S5
S5S5S5S5S5S5S5S5S5S5S5S5S5S5S5S5S5
S5S5S5S5S5S5S5S5S5S5S5S5S5S5S5S5S5S5S5S5S5

Now imagine going through a thirty-minute presentation in a large ballroom, staring at a screen with small fonts and poor background contrast. I'd give up after the first few minutes, and you probably would too.

Font size

Many presenters use text that is just too small to read. When authoring a slide deck, you need to be aware that reading off a projection screen or flat panel TV is a completely different experience from reading off a printed page. The visual dynamics—viewing distance, size perception, even the fact that one uses transmitted light as opposed to reflected light—all affect the experience.

Simply put, if something is illegible on screen, why bother putting it there?

In case you're unfamiliar with it, the unit of measure for font size is points. A point is 1/72 of an inch. While the system was originally designed for print, it just carried over through the years to digital. You also need to keep in mind that while the point system is used across all modern, digital fonts, this doesn't necessarily mean that a certain point size of a particular typeface, will look roughly the same size in another. Here's an extreme example:

20 pts Arial Black

20 pts Carpenter Script

You'll know you've made the right font choices when a person seated at the very back of the room can read even the smallest text in your presentation without squinting.

I won't go into the long explanation of why that is, but it has to do with the fact that different fonts have different 'font metrics.' It's enough that you're aware of it. Here are 14-point versions of some popular, cross-platform fonts:

14 points Calibri Regular

14 points Trebuchet MS Regular

14 points Arial Regular

14 points Georgia Regular

14 points Franklin Gothic Book

14 points Times New Roman Regular

14 points Verdana Regular

14 points Baskerville Regular

Again, notice how some typefaces appear 'bigger' than others, even at the same point size. For the purposes of this book, I'll use Calibri as a reference point for relative point sizes.

So, you might be wondering, what font size should I use to make sure what's projected on screen is readable to the average person? The engineer in you might say "that depends on a whole list of variables!" Screen size, room lighting, the projector's brightness rating (in lumens), font type, font color, background color, a person's distance from the screen and how good his eyesight is. The engineer in you would be right. It really depends on a lot of variables—some easier to control than others.

Since there's no single perfect answer, let me share with you a few "rules of thumb" on font sizes that have always worked for me.

For small boardrooms, I would use a font no smaller than 20-point Calibri. For large ballrooms with large audiences, I'd set that minimum to at least 24 points, assuming

the organizers of an event had the good sense to use a proportionately larger screen with a more powerful projector to match.

While it's reasonable to assume that most of the people that will be picking up this book will be engineers, it's possible there may be some amateur layout artists and graphic designers in the mix. I can almost hear your collective groans of protest at my font size recommendations. 20? 24? That's too big!

Who's in control?

Trust me—slide deck layouts are a totally different animal from printed or even web page layouts. To begin with, viewers of printed and web pages have complete control over page turns, which are based on an individual's reading pace—fast or slow—or reading style—skimming vs. scanning. In the case of web pages, all browsers give users the ability to increase font sizes with a few clicks, anyway.

In a presentation, you are the one in control, and for good reason. You want to be the one managing flow and pace, because your audience has to connect the audible and visual information you're dishing out. You have to be in control because you're the one having to deliver a message within the time limit you were given. By the same token, you have the responsibility for making sure everyone can read what's on your slides.

Using large font sizes will also force you to simplify your tables.

Have you ever been presented with a 20-row 10-column table of numbers on the screen? The presenter then says, "Please focus on the number on the seventh row, third column."

Seriously? Can you imagine the audience counting seven rows down then three columns to the right? I might humor the presenter the first time if I'm really interested but no more. It's too much effort. Another perfect example of self-inflicted audience fatigue.

The big lesson to learn here is to make a solid attempt at condensing the information on screen—limiting it only to what you have time to talk about. A large font size forces you to have fewer rows and fewer columns in a table. It reminds you to condense and limit the information to only the most relevant.

Larger fonts also force you to not only reduce the number of bullets but also produce shorter, more thoughtfully condensed messages. While you're at it, when constructing bullet points, distill them to just key words, not complete sentences. It's your job as a presenter to complete the thought. If you can do away with bullet points altogether—bravo!

Large fonts. Simpler tables. Fewer, shorter bullets. Easier on the eyes.

Good contrast

Let's focus on the very important topic of contrast. This is the perceived difference in brightness, color or texture between an object and its underlying background. Some presenters make it difficult on their audience's eyes by placing text against backgrounds that don't provide sufficient contrast. Poor contrast only gets worse with bright ambient room lighting.

We can have a long discussion about the color wheel and how people perceive color, but we won't. For our purposes, let's just condense it to five rules of thumb.

1. **Garbage in, garbage out.** How your slides look on your laptop or computer screen is as good as it gets. If the contrast on your computer screen is iffy, it will look just as iffy if not way worse on a projection screen. Hedge your bets on things looking worse. Moreover, many projectors or flat screen TVs may not be color-calibrated the same way as your computer. What looks yellow on your computer might look orange on another. A 20% color tint on your laptop might just look white when projected. Prepare for disappointment. When push comes to shove, go with **black on white**—or **white on black**—the two polar opposites when it comes to contrast.

2. **Don't put similar colors side-by-side.** There are certain color combinations to avoid, because their relative luminance levels are too close to each other. You might be a big fan of monochromatic color schemes (it's actually an option in many presentation apps) but unless you're able to assign dramatically different luminance/brightness levels to each then it's hard to distinguish one color from the other.

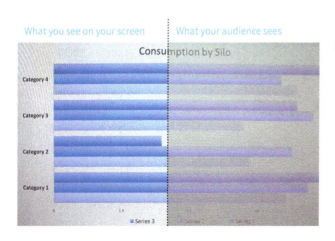

✔ 15% against 80% blue

✔ 90% against 10% red

✔ 100% against 15% grey

✔ 10% against 100% green

✖ 10% against 100% yellow

✖ 100% against 20% yellow

Be yourself; everyone else is already taken

Be yourself; everyone else is already taken

Be yourself; everyone else is already taken

Be yourself; everyone else is already taken

3. **Pale against dark—dark against pale.** When color is important (such as when you're required to use a company template with a prescribed corporate color pallet), try juxtaposing very light values with very dark values whether you're using tints (monochromatic) or going multicolored (polychromatic). The operative word here is *very*. Put very pale on very dark, or the reverse. When using monochromatics, combine a 10% tint with 80% or 20% against 100%—go for a minimum of 70% difference between light and dark, and you will likely achieve good contrast. On the left are some examples of tint combinations that contrast well.

When using monochromatic schemes, avoid naturally bright colors like yellow or bright orange as you won't get much contrast as in the last two samples on the left.

4. **Avoid overlapping "pure" and highly saturated colors on top of each other.** This causes a phenomenon known as *color scintillation*—a bothersome or irritating sensation one gets that can even cause headaches with certain individuals. It's this feeling you get in your eyes when two colors feel like they're 'vibrating' against each other. This feeling is intensified when the material is viewed using a computer monitor or flat screen TV that uses transmitted light, which is inherently brighter. Here are some headache-inducing examples on the left.

5. **Consider the color-blind.** One in 12 men and one in 200 women suffer from one of three major types of colorblindness. That's 4.5% of the population. Depending on the type of color blindness, most will have difficulty distinguishing between certain colors, most commonly red and green.

Knowing these 5 rules of thumb should cause us to rethink or revisit some popular visual devices we've used in the past, and how we execute them. For example, the use of "traffic lights" is a common practice when illustrating scenarios where we have to stop, go, or proceed with caution. Showing differences in both color and form will help.

Instead of this...   try adding words...   or universal symbols.   Because this is what a colorblind person will see:

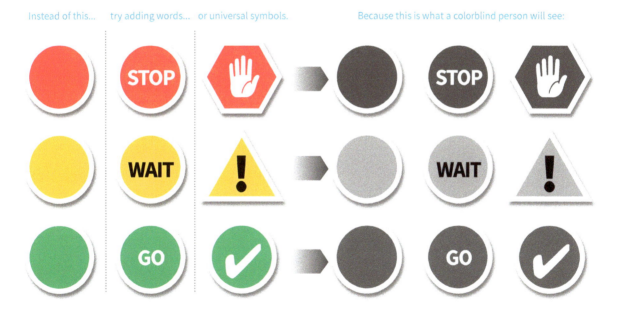

By depending more on shapes and universal meanings attached to certain shapes and symbols instead of just color, we remove as much ambiguity and confusion as possible. Remember, our goal here is to make messages as easy-to-understand as possible, by as many people as possible.

Another instance where we need to be aware of the possibility of color blindness getting in the way of understanding is when objects are right next to each other and the only thing making them distinct from one another is color.

A perfect example is when we assign colors to slices in a pie chart. Here's an example of an improperly colored chart, and how a colorblind person will perceive it.

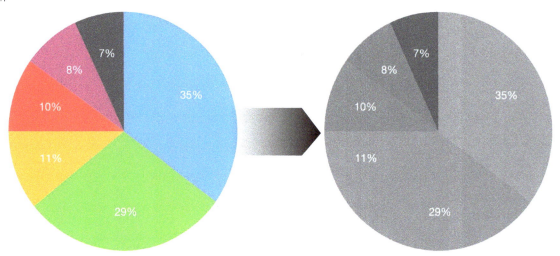

Above is a pie chart we might use in a typical presentation.

A colorblind person would have a hard time telling apart the slices of the pie chart.

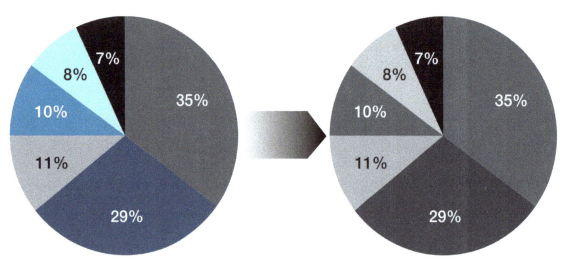

By applying what we know from Rules #2 and #3, we can make this chart readable to anyone.

Notice how the contrasting tint values translate well in any 'color situation'.

Notice how, in addition to alternating very pale and very dark versions of colors to adjacent slices, we've applied the same rule to the overlying text—white text on dark values, black text on pale. We've also made the numbers bigger for easier viewing.

Why bother for a small minority in the audience? Well, what if the person who gets to decide on your proposal is colorblind and couldn't fully grasp the message in your slides because of your use of colors? One in 12 men? I don't like the odds.

We're not diving any deeper into the topic of bad color combinations that print poorly on black and white printers. Some lessons are best learned the hard way.

Noisy slides

Noise refers to any unwanted sound that interferes with what you want to listen to. In the field of communication, it is defined as any irrelevant, redundant or superfluous information that gets mixed in with the useful or important. Safe to say, noise is unwanted and oftentimes annoying.

What does an engineer do when a desired signal is embedded in noise? Between us engineers, we raise the signal-to-noise ratio by filtering out the noise and amplifying the signal.

In plain English, we eliminate as much noise as possible to increase the visibility of the desired information. The lesser the noise, the easier it is to see (and understand!) the useful information.

Can you spot the noise?

Modular ElectroCoagulation Process Replaces Primary Lime Softening or MVC Evaporators

DGF-Deoiled Produced Water

Constituent	Units	Deoiled Produced Water	Permeate Quality (filtered)	% Change
pH	Standard Units	7.4 - 8.2	8.0 - 8.5	–
Total Sulfides	mg/l	1.5	<.3	-80%
Calcium	mg/l	6	2	-67%
Magnesium	mg/l	2	1	-50%
Sodium	mg/l	2000	1900	-5%
Potassium	mg/l	20	19	-5%
Chloride	mg/l	3900	3800	-3%
Sulfate	mg/l	130	120	-8%
TDS	mg/l	6100	5700	-7%
Total Alkalinity	mg/l as $CaCO_3$	600	500	-17%
Total Hardness	mg/l as $CaCO_3$	23.2	9.1	-61%
Silica (as SiO2)	mg/l	220	10	-95%
Water Soluble Organics (TOC)	mg/l	<500	<150	-70%
Total Oil and Grease (after DGF)	mg/l	<10	<2	-80%

Before   After

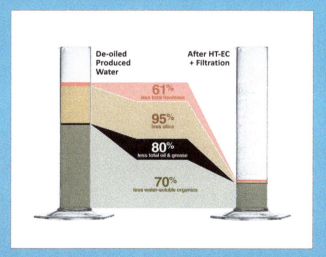

The original slide on the left shows a complete chemical breakdown of all the compounds before and after an applied process—details that belong more on a printed report. On a slide, it's noise.

Turns out, only 4 items on the list were important and the changes were magnified for clarity. The photo of the vials, the other data, even the huge title bar were all dispensable.

Can you spot the noise? *continued*

Swooshes and repeating borders with logos on *every single slide* are not only tiresome, they eat up precious screen real estate. Adding graphics and photos that don't add value or strengthen the message are equally useless.

Here's the same slide sans the noise. Without useless repeating borders, every slide can look fresh and new. Using a relevant photo as a background makes the experience more immersive and impactful.

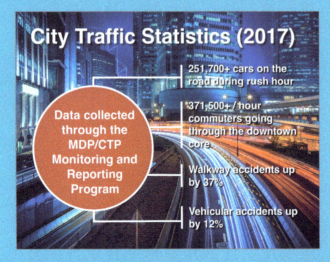

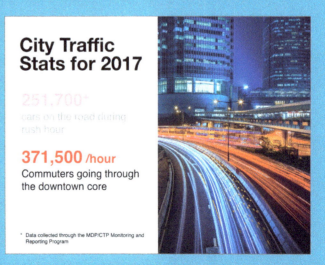

While the background night shot is stunning and even relevant to the message, there's just way too much detail in it that it gets in the way of readability. Just watch your audience squint at this one!

Here, we've managed to keep the high-impact photo, but at the same time bring out the information using a combination of larger fonts, and a controlled reveal, focusing on only one message at a time.

Some people place objects on a slide that are either irrelevant (does not help with the message or add value) or distracting (interferes with the message). These objects are, for all intents and purposes, noise. So I call these "noisy slides."

Why would people add noise to their slides?

Quite possibly, the company you work for requires you to use a template with a background design, a logo or graphic border on top with a matching footer on every single slide. This outdated practice is unfortunately still quite common with a lot of companies. The repeating elements make for wasted screen real estate—real estate you can be using for something important like actual messaging, or precious white space that helps create more focus. If you find yourself in this unfortunate situation, perhaps it's time to have a conversation with your branding and communications people.

While a picture is, indeed, worth a thousand words, for presentations we need to qualify that further. A relevant, well-chosen picture is worth a thousand words. Otherwise, it, too, is noise. Many authors get uncomfortable with empty space and feel they have to plug it with a picture or graphic of some kind, without determining objectively if it helps instead of distracts.

And while we're at it, multiple pictures on one slide seldom work. Even if they seem relevant, there may be too many objects competing for attention. One well-chosen picture that helps with the message should be plenty for one slide. It could be because the presenter felt that adding an icon or clip art here and there looks cool. Again, if it helps with the messaging, sure. If not, it's noise. Multiple icons, probably not helpful.

Look at the objects on each of your slides. Does each object help with the one message the slide is supposed to have? If so, great. If not, it's noise. Even if it makes the slide colorful and pretty but doesn't help with the messaging, it's still noise.

Noise in your slides is irrelevant and distracting. Take it out.

CHAPTER 8
Easy on the mind

It's one thing to make the message easy to understand by creating visuals that are easy on the eyes. It's another thing to make it easy on the mind. Make it easy for your message to easily flow into your audience's mind.

The key to making anything easier on the mind is by simplifying, and there are a number of ways to achieve simplicity.

In algebra for example, we might say:

$$xy + 3y = x^2 + 4x + 3$$

But factoring out $(x + 3)$ from both sides of the equation, we end up with just:

$$y = x + 1$$

We simplified the equation by canceling out what's common or irrelevant. I'm sure you've done similar exercises thousands of times in college.

Simplifying is often necessary to make things understandable, manageable and ultimately, useful. But simplifying takes effort—sometimes, a lot of effort. It's not as easy for some people as it is for others and, more often than not, what makes it difficult is our old friend, the 'curse of knowledge'—remember him?

Explaining technical messages to a non-technical audience often requires a ton of simplification. I've seen some of the most complex concepts made understandable—even enjoyable—because someone took the time and effort to make it so. We can simplify without being simplistic and without dumbing down the message. It's like the difference between someone being child-like as opposed to childISH.

So the question is, how do we simplify?

Tech talk

I once tried to read a graduate-level social science technical paper written in English. It was so technical, I understood absolutely nothing. Zip. I consider myself a pretty smart fellow but this time, I drew a complete blank. It was like trying to read something written in a foreign language. People in that field might understand it, but I certainly couldn't.

This was an extreme case. The language and style used in that paper were probably meant for people in that field and at that level. There would be no knowledge gap between the material and the intended audience.

But, as a technical person faced with the task of addressing a non-technical audience, a knowledge gap exists to begin with. In some cases, it can feel like a yawning canyon—like the one I faced trying to understand that social science paper. There's no skirting the issue. Your intended audience lives on the opposite side of that canyon, and it's your job to build the bridge, not theirs.

In other words, when addressing a non-technical audience, there is no room for jargon—or as I like to call it, "tech talk." Sounds like an old clock, doesn't it?

It reminds me of a time I was presenting and at one point, I got too comfortable and unwittingly let loose with a string of technical terms that sounded perfectly normal to me, but which brought a sudden silence to the room. The silence was punctuated by the blank stares from the audience, along with the sound of the wall clock mocking me—tech…talk…tech…talk…tech…talk…

One way to make something understandable is by avoiding tech talk and reinterpreting the message in simple, everyday terms that people can relate to in real life. For example, if you were to start a talk about a fictional invention with:

> " *I'm here to tell you about a 50-volt 9,010 ampere-hour battery that fits in a 550 cubic-inch container. Isn't that amazing?*"

You would probably be greeted with blank stares and questions like, "what are ampere-hours?" They probably wouldn't know how big a 550 cubic-inch box is. Even engineers would need to do some computing before they could imagine the box dimensions. I know I would. You must make those numbers meaningful to your audience. You might not even have to actually mention the numbers.

Drop the tech talk altogether and just say:

> " *I've invented a super battery that can power your entire home for 2 weeks without recharging. And it is small enough to fit in a shoe box! Amazing, right?*"

I'm sure they'd agree. Notice the complete absence of jargon, particularly how you've added context to the invention by couching it in terms that are both easy to understand and applicable to daily life.

Tech talk can take other forms. Engineers are used to automatically converting numerical unit prefixes like "kilo" for thousand, "mega" for million and "giga" for billion. We take these things for granted. For a non-technical audience, it's hardly automatic.

For example, some people might understand "kilowatt-hours." Some may even imagine ten 100-watt light bulbs lit for an hour. The term is common enough. But it becomes less understandable when we say megawatt-hours. Instead of saying 10 megawatt-hours, just say 10,000 kilowatt-hours to avoid the need for mental conversion altogether.

Another way is to just avoid using numerical scales altogether. Instead of saying "This wire has a diameter of 50 micrometers", we could simply say, "This wire is as thin as human hair." Nothing beats real-world comparisons for quick understanding.

A quick way to check your jargon level is by having a non-technical friend scan through your slides. If there are words he or she doesn't recognize, understand or normally use, consider substituting them with simpler or plain language.

Steer clear of faddish buzzwords you would never use in day-to-day conversations. And it would do you well to avoid clichés while you're at it.

"... Moving forward, we can really move the needle and take this to the next level by leveraging our differentiators, maximizing our limited bandwidth and making sure we get all the low-hanging fruit while thinking outside the box..."

Another thing to watch out for with non-technical audiences are acronyms.

" *You've been cc'd the PDF of the SOQ containing our EOR, MOS and THO capabilities like SAGD and CSS.*"

It's one thing to err out of ignorance, but if you're one of those folks who secretly think that tech talk creates the perception of higher intelligence or superiority, please—get over it. The only perception you create by hanging on to jargon will be that of being inconsiderate. If you genuinely want to succeed and be understood, then you need to speak plainly.

Analogies

You could be in a situation where you and your audience are speaking the same language but simply aren't connecting. You may just as well be speaking different languages. You need a translator.

When people with very different backgrounds and training need to connect, they need a bridge of some kind—a common ground or translator of sorts to explain something discipline-specific to the other. Try explaining something electrical engineering-specific to a medical doctor. Good luck.

This is where the use of analogies can be your greatest ally. You've probably heard how water can be used to explain the concepts of voltage, current and resistance. I've used that analogy lots of times in class but this time, let me use something closer to my heart—beer— to explain the fairly complex concept of "power factor."

I once worked as an engineer in a medical facility. The building administrator was a medical doctor. He saw the building's electric bill and noticed a large penalty for having a low power factor. He got advice to install power factor correction capacitors but didn't understand what power factor was.

My boss asked me to explain "power factor" to the doctor. I thought about it and drew a mug of beer with a lot of froth. I then said:

" *Imagine ordering a mug of draft beer. You always get some froth with your beer. You want less froth to get more beer but you pay for a full mug of beer anyway.*

So you tell the bartender, 'Please pour slowly, with the mug slightly tilted. I'd like less froth and more beer.'

With electric power, you have something similar going on. You always get reactive power with real power. Real power is your beer; reactive power is the froth. You only really need real power. You don't need reactive power but you get and pay for the reactive power anyway.

Having motors in the building—air conditioners, compressors, elevators—increases the reactive power. You get more froth.

Power factor correction capacitors work like the slow pour with the tilted mug—you reduce reactive power, the froth, and get more real power, the beer."

Electrical engineers might argue, well that's not exactly how it works and they would be right. To be totally accurate, I would have to touch on vectors, phase shifts between voltage and current, load impedance and all that. Problem is, the technically accurate explanation would be lost on the doctor. It wouldn't help.

As analogies go, the mug of beer allowed me to explain a fairly technical concept to a medical doctor. He certainly didn't want to continue paying for "froth." This was enough to convince him to contract a company to install power factor correction capacitors and understand in general terms why he was doing so.

Bite-sized messages

What would happen if you tried to swallow a plateful of food in one go? You can't. And if you tried, you'd probably choke.

So how does one consume a plateful of food? One small, chewable bite at a time.

So how do you present a slide full of bullet points? One bullet at a time. Makes sense?

If you have a slide with five bullets, presenting all five at the same time would be like forcing your audience to swallow everything in one go. Sad thing is, they'll actually try. Most, if not all of the people present will start reading ahead of you, and once they do that, they momentarily disconnect with what you're saying. This disconnection is, for all intents and purposes, choking.

What I do is use PowerPoint animation to make the bullets appear one at a time, guiding the audience's eyes to the latest bullet. I call it a **controlled reveal**. This makes the audience quickly read the current bullet while I talk to it. I then click my way to the next bullet and the next until I've built up the slide, completing the "thought set."

Reveal your bullets one at a time, guiding your audience's eyes to the latest bullet while you talk to it.

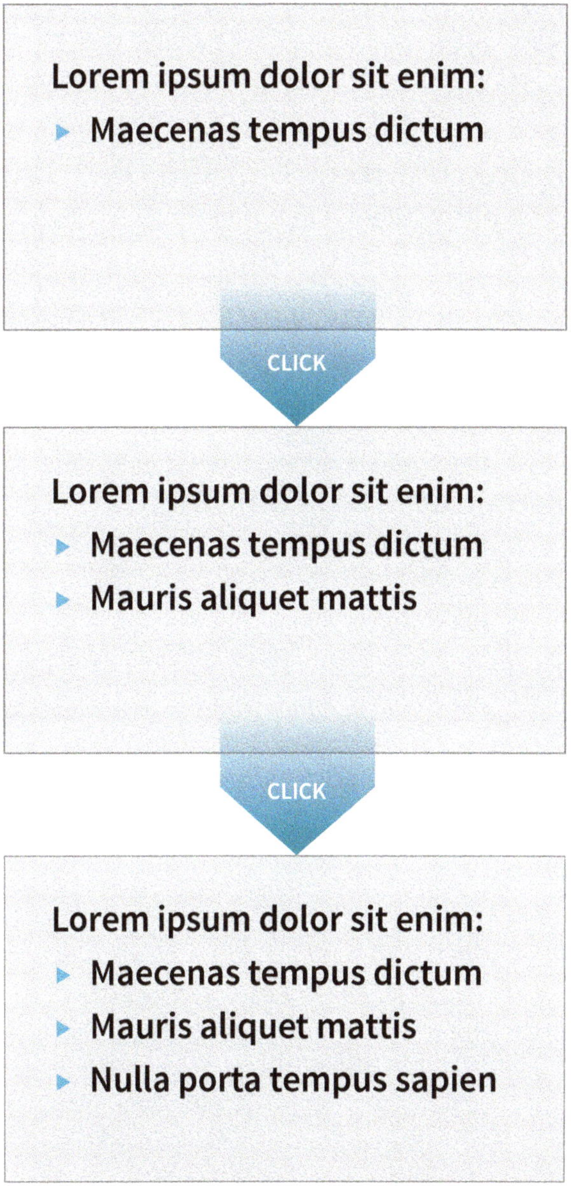

Even better, dim the bullets that you're finished with, to direct focus even more.

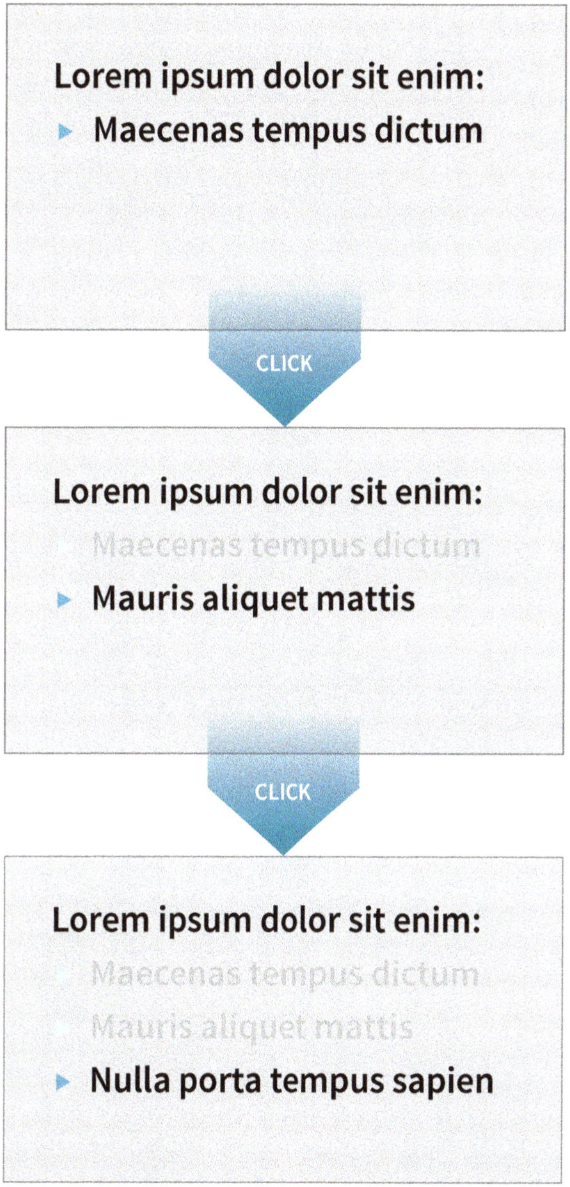

Lorem ipsum dolor sit enim:
▶ **Maecenas tempus dictum**

CLICK

Lorem ipsum dolor sit enim:
▶ **Maecenas tempus dictum**
▶ **Mauris aliquet mattis**

CLICK

Lorem ipsum dolor sit enim:
▶ **Maecenas tempus dictum**
▶ **Mauris aliquet mattis**
▶ **Nulla porta tempus sapien**

Lorem ipsum dolor sit enim:
▶ **Maecenas tempus dictum**

CLICK

Lorem ipsum dolor sit enim:
▶ Maecenas tempus dictum
▶ **Mauris aliquet mattis**

CLICK

Lorem ipsum dolor sit enim:
▶ Maecenas tempus dictum
▶ Mauris aliquet mattis
▶ **Nulla porta tempus sapien**

Do the same thing when presenting a process or a procedure. Break it down into steps by doing a controlled reveal. Use animation to manually flash each step one at a time. Build the slide as you go along.

The more complex the system, the more you should make an effort to break it down into bite-sized messages that the audience can digest. That is your role as presenter.

While we're on the subject of bullet points, you've probably noticed by now that I'm not a big fan of them. But if you absolutely must have them, make an effort to condense each bullet to a phrase or even a word. Make the bullet a shorthand version of what you're about to say. As much as possible, avoid using complete sentences in bullets. That creates distraction. As you flash each phrase or word, you talk about it in complete thoughts and sentences. This way, the bullet points become reduced to little cues or triggers that are far easier to remember.

Here's a concrete example.

An agency issues ID cards to over 30 million of their members. It had a complicated system that included biometric capture stations at many locations, an automated fingerprint identification system, an ID card production facility and a mailing system.

The system was more than ten years old and was performing way below capacity with frequent breakdowns. An upgrade was way overdue.

The board needed to understand the problem. I did the presentation.

My approach was to break the systems down into conceptual blocks. First, explain the 4-step process, starting from a member requesting for an ID card, to receiving it in the mail, then describe how each of those

steps has a corresponding subsystem of computers and equipment. Four process steps, four subsystems.

Explaining the process was easy. The process slide was built up by making a process box appear one at a time.

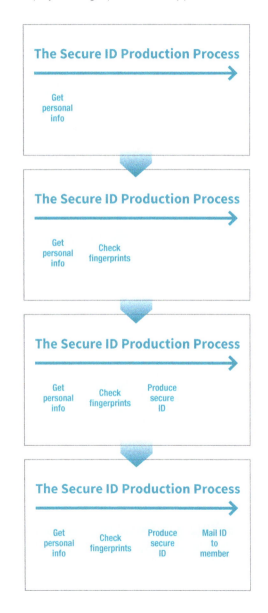

Once they grasped the process, understanding the corresponding subsystems was a piece of cake. The board was able to clearly visualize the system as having four subsystems, each responsible for each of the process steps.

The system slide was also built up by making an icon for each subsystem appear one at a time. Notice how I was able to avoid conventional bullet points and resorted to using icons instead.

By controlling the flow of information visually and in stepwise fashion, I was able to hold the board's attention as I carefully described each subsystem's shortcomings and how they affected the steps that followed.

At the end of the presentation, the chairman commented, "This is the first time we've come to clearly understand the current system and the need for a new one."

As I left the room he added, "And we hope to see more of your presentations in the future."

Sorry, but I just had to share that last comment from the chairman. It was a yahoo moment for me.

This is an example of a technically complex presentation that was successfully understood and appreciated by a non-technical audience.

The move towards bite-sized messages doesn't always have to be a grand effort over a large complex concept. Make it a habit, even for smaller systems.

On a smaller scale, here are few more tips for delivering bite-sized messages. Many of these are sprinkled throughout the book.

1. Using 24 points or larger for text forces you to have less text on a slide and fewer numbers in a table. Instead of spreading a buffet of messages on one slide, go for just a plateful dished out a bite at a time.

2. Then there's focusing on the number "ONE." Stick with just ONE message per slide. If a slide has two messages, split it into two slides. Slides are free.

3. If you have charts, keep it at ONE chart per slide. The chart must be simple, with only ONE message per chart.

ONE bite-sized message at a time, delivered in a logical sequence will help your audience understand your logic.

Guide the eyes—guide the mind

Eye-tracking studies tell us where a person's eyes focus on a page or a slide. Website designers use these studies to determine what draws people's attention on a page, and in what order or sequence. But we won't get into that.

Ideally, you want to have the ability to direct your audience's attention so they're looking at the right thing at the right time. You can achieve this by designing your slides carefully, controlling the size and look of elements, when they appear, and how all the elements come together nicely to create a cohesive thought.

Three quick ways to pull attention to a word or phrase is by making it either **bigger**, **bolder** or changing its **color** to contrast it with the rest of the words. While red seems like an obvious choice, you may need to use colors from a prescribed palette if you're using a company template.

Three quick ways to pull attention to a word or phrase

Make it bigger

> The only thing worse than being blind is **having sight** but **no vision**.
> —*Helen Keller*

or

> The only thing worse than being blind is **having sight** but **no vision**.
> —*Helen Keller*

Make it **bolder**

> The only thing worse than being blind is **having sight** but **no vision**.
> —*Helen Keller*

or

> The only thing worse than being blind is **having sight** but **no vision**.
> —*Helen Keller*

or change its color.

> **The only thing worse than being blind is having sight but no vision.**
> —*Helen Keller*

or

> **The only thing worse than being blind is having sight but no vision.**
> —*Helen Keller*

If you want to highlight a cell or figure using an arrow, create one using the shape tool but leave the fill in. Rotate and position it in a way that works for you then apply an entrance animation. You can keep it simple by using an **Appear** or **Fade In** entrance. If you want to get a bit fancier, you can use **Fly In** or **Wipe**. I recommend you adjust the **Wipe** or **Fly In** direction to start from where the tail of the arrow is, so an arrow pointing up, should **Fly In** or **Wipe** from the Bottom, and one pointing left should start the animation from the right, and so on.

Learn how to use these animation features—it's a useful way to direct your audience's attention.

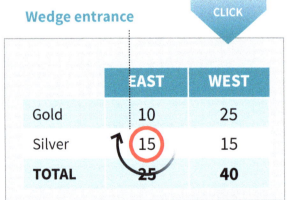

Precision distraction

Engineers love to be precise with their numbers. For some, it's a need they just can't shake.

Let's say for example, you wanted to report that a company's electrical consumption in a year was 1,702,365 kilowatt-hours, which cost the company $190,154.12. While precision is important when making an analysis or writing a technical report, we need to make some adjustments when presenting to a non-technical audience. With the latter, precision can often be a distraction rather than helping the message along.

Think of it as mathematical 'paraphrasing'. The figures above can be more quickly understood as 1.7 million kilowatt-hours, at a cost of $190 thousand. For a non-technical audience, any more precision doesn't add value to the message.

If anyone were to request for the work behind the numbers, I'd give them the spreadsheet in a separate, hard or soft copy.

CHAPTER 9

Easy Charts

Engineers are comfortable with charts. In fact, technical presentations by engineers are often peppered with them. They're a great way to visualize data and communicate certain complex messages.

But we engineers need to understand (and accept) that the way we look at charts often differs from the way non-engineers see them. Especially in a presentation where understanding has to be quick, we need to surface the message of a chart in ways that will generate that quick understanding. My personal rule of thumb—your audience has to be able to grasp the main idea behind the chart in 3 seconds or less. We'll call this the "3-second rule."

In under three seconds, your audience should see an obvious trend, a clear majority, discrepancy or disparity in a group within a data set. Your spoken explanation can 'fill in the blanks' with regard to the details of, say, what the trend implies, what may have caused it, how the data was gathered, and so on.

Here are four suggestions to help you meet the 3-second rule:

1. Make sure you're using the right *kind* of chart.

2. Set a limit of one chart per slide.

3. Set a limit of one message per chart.

4. Make the chart clean and simple.

I presented these same suggestions to a group of managers once. One of them, an engineer, remarked, "I think we all know these rules but we just forget to use them."

It's important to know how to choose the right kind of chart for the data to send the right message. Let's look at the more common ones.

Cartesian

Engineers are most familiar with the **Cartesian** or **X-Y Chart**, where we plot the relationship between one variable, **x**, and a second variable, **y**. In essence, it represents **y** as a function of **x** or **y = f(x)**.

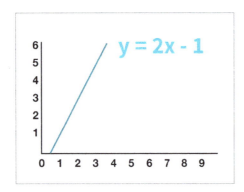

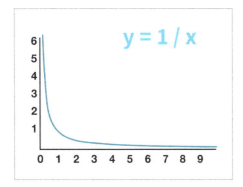

X-Y charts are often used to express data as a function of time, which is commonly placed on the X-axis. The charts below have Y-variables that vary across time.

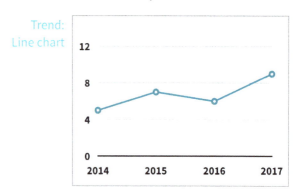

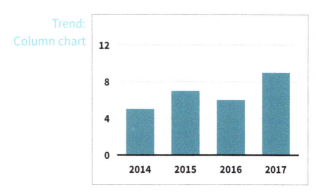

While a column chart can also show trends over time, I still find a line chart simpler to understand.

Line, Column and Bar

Line and **column** charts are sometimes combined to overlap two time trends. In this example, the column chart was the **target** data series, while the line chart was the **actual** data series. The **actual** data in the next graph was also presented in red to draw attention to it.

Mixed trend
chart

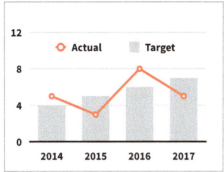

But for charts with long, descriptive titles along an axis, I recommend using a bar chart instead of a column chart. This will avoid having your audience tilt their heads to read axis labels.

Column (vertical) and **bar** (horizontal) charts are often used to compare groups. The data for the bar chart below can be presented using a column chart just as well.

Instead of

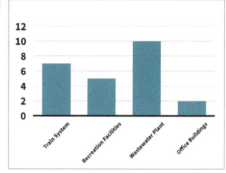

as a column
chart

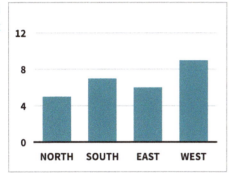

...do this.

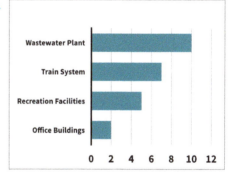

as a bar
chart

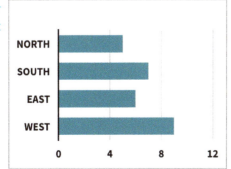

Notice that I also sorted the data from largest to smallest value before creating the bar chart. That extra step of sorting helps surface an additional insight beyond just dumping the data on your audience.

This saves your audience the effort of having to sort the bars in their heads and figuring out that insight.

Gantt charts

Another popular type of bar chart is for showing a project schedule, highlighting when certain stages of a project start, end and overlap with one another—in other words, a Gantt chart.

Most people will use a spreadsheet program to create Gantt charts, because of the available templates. But the formatting on these templates leave much to be desired and are usually meant for print media (e.g., 11 x 17 sheet). The details on the axes tend to be small, bordering on inscrutable. This level of detail might be okay for print, but they would look terrible on a projected presentation slide. The chart below is a typical example:

As an exercise in readability, try propping up this page then back up about five feet. How readable is it?

In the same way we've improved readability and understanding on the previous chart examples by condensing information and removing unnecessary 'noise', I'm going to show you how to make the same information readable, even from a distance.

One of the first types of noise to hunt down and squash is repeating information. Notice how the dates on the graph below repeat months and years? Use the **Merge** cell command and group together headers for dates that fall within the same month, and months that fall within the same year.

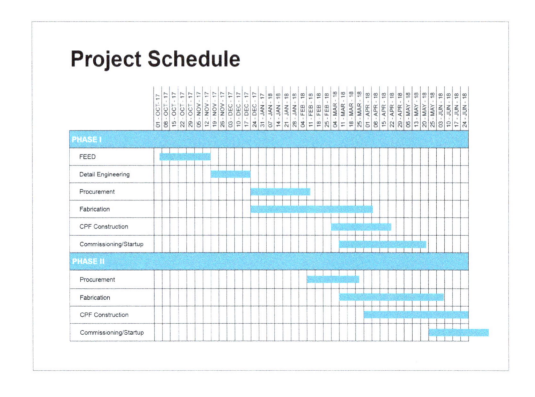

If the schedule information is such that it's relevant to show how tasks cross from quarter to quarter, then by all means, group the months by quarters, condensing the information even more.

If certain milestone dates need to be highlighted, it's easy enough to create a graphical pointer (e.g., a tiny red triangle) with a defining label or call-out.

Another way to improve visibility is by minimizing the gaps between bars, and eliminating secondary grid lines on the Y-axis. If you feel that the secondary grid lines are important, at least make the primary grid lines bolder in comparison, so your viewers can easily tell when, say, a quarter ends and the next one begins.

If the tasks are too many (more than 10), then you might consider either splitting the chart vertically across multiple slides, or scrolling it upward using animation. Either way, try your best to limit what you show on screen to what you can immediately talk to. Don't distract your audience with too much information that will cause them to tune you out and start reading ahead.

Prop up the chart below and compare its readability from five feet away to the one on the opposite page. Can you see the difference? By making the extra effort tweaking how much information is shown, how it's grouped, allowing for larger fonts and removing unnecessary lines and repetitive information, you can help your audience see the information more clearly and quickly.

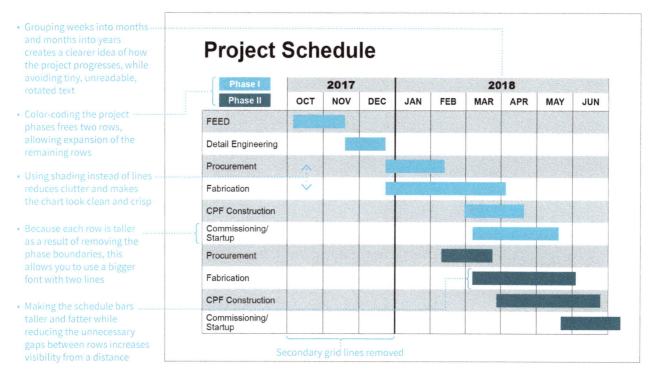

- Grouping weeks into months and months into years creates a clearer idea of how the project progresses, while avoiding tiny, unreadable, rotated text

- Color-coding the project phases frees two rows, allowing expansion of the remaining rows

- Using shading instead of lines reduces clutter and makes the chart look clean and crisp

- Because each row is taller as a result of removing the phase boundaries, this allows you to use a bigger font with two lines

- Making the schedule bars taller and fatter while reducing the unnecessary gaps between rows increases visibility from a distance

* Gantt charts are the one exception that breaks the 20-pt.-text minimum rule given the number of labels. Even so, we were able to manage with a minimum of 14 pts on the chart above.

Pie charts

Now for the consistent favorite—pie charts. When used to surface the right message for a given data set, pie charts are great. People seem to just love pie charts, often making it the chart of choice even when it's not the best chart to use for a given message.

A pie chart compares portions of a whole and works best when demonstrating proportion. This is perfect when you're trying to make the argument around how important (or insignificant) one portion means to the whole.

For example, the first chart on the right says that a particular city's consumption of energy is more-or-less evenly split between electricity, natural gas and diesel. That's a good insight and the use of a pie chart works. The second chart says that a majority of energy costs are in electricity. The use of a pie chart also works, especially when we use a distinctly different color from the rest of the slices such as red to make a key slice jump out, especially if it's a key talking point. In both examples, the use of a pie chart is effective for the intended insights.

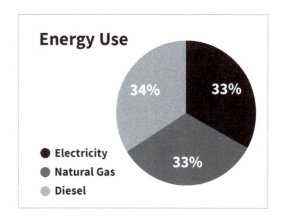

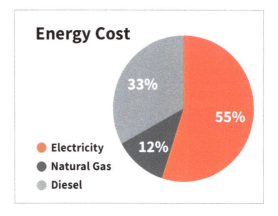

Some data sets, on the other hand, just don't work using pie charts. Data that trends with time, for instance, are best shown as a line chart:

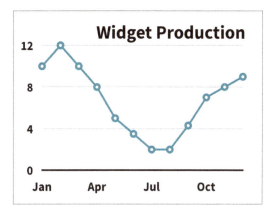

You could show the same data on a pie chart, but it wouldn't make much sense:

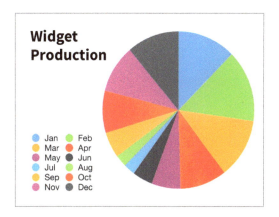

Pie charts fail when the message requires the audience to **distinguish between closely-valued proportions**. For example, In the pie chart below, Product A leads with 33% of production, while Product C is a close second at 31%. Do you see Product A as larger than Product C? Without labels, that isn't obvious in the pie chart, is it? The use of a dark color with Product A and a bright color with Product C might even make Product C appear larger.

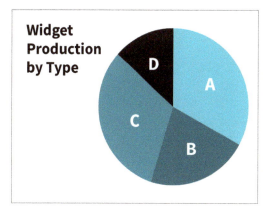

You might want to use a **data-sorted** column chart instead. That would make the data much clearer:

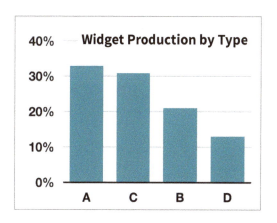

Pie charts (continued)

Pie charts also fail when you try to put in too many slices, or when you fill the chart with too many labels. This would be akin to having too many bullets, too many messages, too many distractions, or too much noise in a slide. That single message you want to show just won't surface.

Line charts, column charts, bar charts, pie charts—these are some of the more common chart types I use in presentations.

With care, and only if the need arises, I would, on occasion, use **scatter**, **stack** and **area** charts. Other chart types however—**radar**, **stock**, **bubble**, **surface** charts—are probably too unfamiliar or complex for non-technical audiences. With these charts, you'd be hard-pressed to satisfy the "3-second rule".

Knowing which chart type to use for certain data sets is the first and most important step in helping you visually demonstrate insights through charts more clearly.

Too many slices

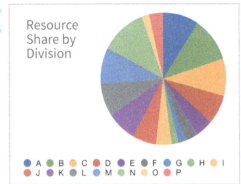

Too many labels

One chart per slide

Multiple charts on a slide are like a cacophony of different messages, each screaming for attention simultaneously. It violates many of our earlier lessons on making a slide easy on the eyes and easy on the mind. Squeezing everything in a limited space shrinks things, making them harder to read.

Sad to say, these slides are more common than you think. This layout might work on a printed page, but it doesn't belong in a presentation slide. So, the issue here is information density—there's just too much for this medium.

To address the situation, don't even consider doing a controlled reveal, fading in one chart at a time—there simply isn't enough space to begin with! The right thing—the only thing to do—is split it up into six slides, one chart per slide. If your idea is to make a compelling argument for what these six charts altogether represent, then I suggest creating a seventh slide that illustrates this fact in a more distilled fashion.

Readability is fundamental. Remember our recommended font size? There's no way you can draw these six charts with 20pt text in the axes. I would challenge any graphic designer to do so, and still make it look pleasing and readable.

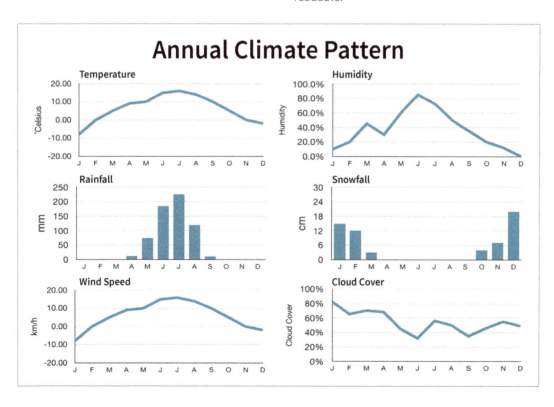

One message per chart

When a lot of data is available, there's always the temptation to cram as much as we can in a single chart. It might look impressive but it doesn't do much by way of helping you deliver a clear message.

Consider the chart below. If each of the seven data sets has its own message, then presenting them simultaneously won't help you meet the 3-second rule. How long do you think it will take the audience to grasp all seven messages? Why not create seven charts, each with one message shown on its own slide?

If the point you want to make is shown by just one data set—in the example below, Facility G—then the other lines are just noise, distractions better left out or at least minimized on the chart.

In the second example, it could be that the point you want to make is that Facility 'G' has been increasing at a faster rate than all the others. That's what we did—the rapid increase in Facility 'G' stands out more, doesn't it?

The minimized lines representing the other Facilities are now just visible enough to provide context, because that's really all you need.

Generally, I try to limit a line chart to just one highlighted data set and completely eliminate the others if context isn't necessary. If you need to highlight two sets for comparison, then all the more you need to remove as much of the 'noise' as possible.

Instead of

...do this.

Clean and simple

In the previous chart, not only did we make Facility "G" stand out, we also made the chart look cleaner, simpler and easier to understand.

Did you notice how we did that?

From a visual/graphical standpoint, we achieved this by:

1. Making the 'G' line thicker, using a bright red color;

2. Making all the other lines much thinner by comparison;

3. Making the grid barely visible to remove background noise, and;

4. Simplifying the Y-axis (multiples of 5 instead of 2) and making the X-axis letters larger.

Here are a couple more examples in making charts look clean and simple.

When using column charts, try to steer clear of fancy 3D charts. A 3D chart may look prettier with all the shadows and subtle highlights but because they're presented in perspective, there's the possibility of distorting the viewer's perception. This distortion gets especially problematic when values are close to each other.

Instead of

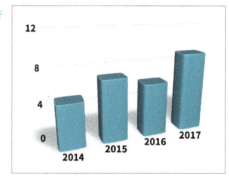

...do this.

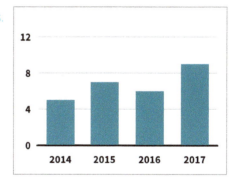

Between the pretty but potentially misleading 3D chart on the top, and the cleaner, simpler and more accurate representation on the bottom, I believe that engineers will go for the latter.

Let me end this chapter with a last word and example on pie charts.

As we mentioned earlier, pie charts work well when showing relative proportions between just a few categories. By few, I mean around six or less. When there are too many pie slices and differences between them become imperceptible, then the message starts to blur into a muddy mess.

The pie chart below shows population contributions by country. Assigning a slice per country results in too much detail while the accompanying legend becomes unwieldy.

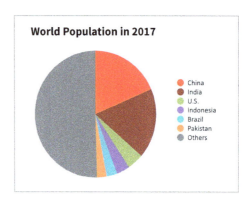

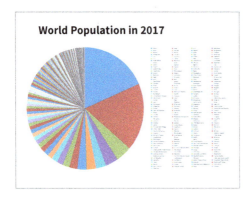

The next chart condenses the information by only showing data from the top six contributors, and grouping the rest into an "Others" category. This delivers the message that half of the world's population live in just six countries.

The third chart below goes further by highlighting the fact that about a third of the world's population live in just two countries.

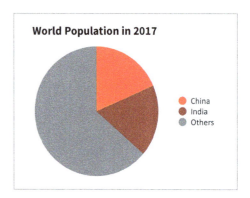

Limit the slices in your charts to allow the important insight to surface. Sum up the smaller slices into an "Others" category.

Of course, the way you distill and group data will depend entirely on what message you want to highlight. The same chart would look very different if countries were grouped by continent, for instance. Either way, visually condensing information will almost always create clarity.

"Everything should be made as simple as possible, but not simpler."

Albert Einstein

10

CHAPTER 10

Typos, inkblots and other horrors

Engineers go to extreme lengths to ensure professional quality work. We wouldn't stand having our engineering work called sloppy, unprofessional, unsafe, inaccurate, imprecise or generally of bad quality.

Shouldn't professional, quality work also show in our presentations as engineers? Is it enough that the numbers and facts in our presentation are correct?

Presentations provide a window into the quality of our work in general. People should be able to assume with relative confidence that our facts are correct, at the very least.

Beyond that minimum, nothing says sloppy more than misspelling, bad grammar, a wrong name or a miscalculated number. I'll be grouping all these kinds of errors under the category of "typos."

What's the big deal? People make mistakes, right? Spur-of-the-moment gaffes like a misspoken word in casual conversation are perfectly understandable. But it's equally understandable for people to assume that you spent a significant amount of time preparing your slides, which makes it a big deal.

If you can't even get your spelling right, how can the audience trust that you got your engineering calculations right? It's called the *halo effect*—the tendency for an impression created in one area to influence opinion in another area. In the audience's subconscious, sloppy spelling must mean sloppy everything else. Worse, if you can't even get the name of your client right, how do they know you're even in the right room, speaking before the right audience? If I were the wrongly-named customer, I would feel very disrespected and react accordingly.

For basketball fans, the story of how Nike lost Stephen Curry to Under Armour is an excellent example of how getting your customer's name wrong can be a deal breaker. Google it.

A typo is like a tiny inkblot on an expensive paper stock. Nobody notices how nice the paper is—all they see is the inkblot.

Okay, so you might be an excellent engineer but a very poor speller. Or maybe English is not your native language so your grammar is weak. This is still no excuse for sloppy slides, and won't stop your audience from casting doubts on everything else.

So, instead of making excuses, let's instead channel our energies toward ways to minimize, if not totally eliminate these non-engineering typos out of our slides.

I would start with a few basic suggestions. These should catch and correct most typos early.

1. **Remember Chapter 2—No Excuses!** You and only you are responsible for your slides—typos, warts and all. Even if somebody else prepared the slides for you.

2. **Set aside time in your preparation** for proofreading and quality checks. Checking for typos and other errors should be integral to your process—not an afterthought at the eleventh hour.

3. **Do a technical accuracy check.** Are your numbers right? Do you know what or who the source is? Is the source credible? Are your units consistent? Metric or Imperial? Celsius or Fahrenheit? Parts-per-thousand or parts-per-million? Kilowatt-hours or megawatt-hours? To an engineer doing a technical presentation, technical accuracy is expected.

4. **Run an automated spelling and grammar checker.** It may not catch all the errors, and you might not agree with everything the tool finds, but it will catch a few obvious errors that you missed.

   A lot of common engineering terms aren't in spell-checking dictionaries. Words like 'deoiling' or 'debottlenecking,' so pay close attention when accepting spell checking suggestions—the most common being split words when they shouldn't be split in the first place.

5. **Read the whole document from cover to cover** even after running an automated spelling and grammar checker. Why? Spell checkers will almost never catch wrong words that are spelled correctly (e.g., complementary vs. complimentary, principal vs. principle). You might also find a few more errors that you, as the author and presenter will see but the tools would not. Here's a common type of mistake when using acronyms: "TIC cost estimate." TIC stands for total installed cost, so "TIC cost estimate" is like saying "total installed cost cost estimate."

6. **Engage an experienced proofreader.** It's not uncommon for people to be blind to typos after having worked on a presentation for so long. Your deck needs fresh eyes that will not only see typos but unnecessary jargon and redundancies. Not everyone is grammatically gifted so try to get help from someone who is.

Then, there should be an effort to get key details right, as these could turn into a big deal if you get it wrong.

7. **Copy names directly from a business card or an official corporate website**, especially for customer and company names. Have a legitimate source you can point to. Find a way to validate further that you have both the right person and the right name by perhaps cross-checking with LinkedIn. Call your contact at that company—all it takes is a phone call.

   Managers don't like it when you misspell their company name. Be aware that some company names have two words separated by a space (Duane Reade, Reckitt Benckiser), some are hyphenated (Daimler-Benz, PMC-Sierra), some have no spaces but instead, have embedded caps (ConocoPhillips, GlaxoSmithKline, ExxonMobil), or both (Bristol-Myers Squibb)!

8. **Be extra wary of accidentally sharing confidential information in your slides.** This breach can come in many forms, from diagrams or spreadsheets that may contain a competitor's name or data, to cost computations that are strictly for internal use. More than just an embarrassment, it can sow distrust and potentially end in disaster.

9. **Be sensitive to regional differences in spelling** (e.g., American vs. British or Canadian English). Center or centre? While or whilst? Aluminum or aluminium? Your audience might see a typo where there isn't one.

10. **Watch out for careless inconsistencies** in spacing, typefaces and font sizes. This usually happens when you cut and paste text from one file to another. The words may be the same, but some people will sense the inconsistencies—the feeling that something's "not quite right." I know I would. It might surprise you how many people will notice when you've inserted a sentence in 10 pts on a page full of 11 pt text. Or mixing a phrase in Arial on a page full of Calibri. Or mixing up different line and paragraph spacings.

11. **Exercise careful brand attribution** when embedding brand names in body text by appending the proper trademark and registration marks to brands, especially when they're not in the mainstream. You need only attribute at the first mention of the brand in a document—no need to keep repeating it.

    There are three commonly used trademark symbols: the superscripted ™, ℠, and the letter R in a circle ®. The TM and SM symbols are used with unregistered marks: TM for trademarks, or marks that represent goods, and SM for service marks, or marks that represent services. The federal registration symbol, or ®, is reserved for marks registered in the U.S. Patent and Trademark Office.

Finally, there's Murphy.

12. **Avoid last-minute tweaks.** These are changes that will no longer have time to go through the suggested quality checks above. Moreover, if the tweak involves additional slides, you won't know how these last-minute changes can affect the flow and timing of the entire presentation because they were never practiced or tested. These are the tweaks that Murphy is just waiting to pounce on.

PART IV

Effective delivery

You have a clear purpose. You know your audience. You've crafted your messages and designed your slides. Now comes the fun part—delivery.

Part 4 is a compilation of lessons on presentation delivery. Consider it your tool kit for effective delivery.

You don't need to read the chapters in sequence. Each chapter will provide a lesson that stands on its own and can help you either way.

CHAPTER 11
Presentation traps

To a large degree, presentations are all about control—controlling the message and how that message gets delivered. You want to make sure you're always on track, on time and most importantly, on point. Lose control, and you run the risk of failing to achieve your objective.

Occasionally, someone in the audience asks a question that catches you off guard. You put your presentation momentarily on hold, try to address it, and before you know it, you've either lost your train of thought or worse, answering that question forced you to change lanes and your entire presentation no longer makes sense.

You had just fallen into a presentation trap. The trap wasn't put there by someone who wanted to trip you. More likely, you unwittingly dug a hole yourself and all it took was one question from the audience for you to fall in.

Message premises

All presentations are built on premises—assumptions, definitions and methodologies. Think of these as one post of a three-legged stool. Kick one leg out and the stool tips over.

If your presentation rests on an assumption, explicit or otherwise, that when questioned is proven wrong, then your entire presentation falls apart. Or perhaps you defined a key term using a non-standard definition. How about using a methodology that an expert in the audience shoots down?

What if two experts in the audience agree that something's not right with your methodology? How long can you continue sitting on a stool with just two legs?

You address the question, get sidetracked, and then get drawn into an argument on the basis for that assumption, the correctness of the definition or the validity of the methodology. You've lost control of your presentation.

One of two things can happen and neither one ends well.

If your responses fail to definitively answer the question and you run out of time, your presentation may get rescheduled for another day, if at all. If your answer fails to erase all doubts but you continue anyway, the unsettled question will weigh against you when you draw your conclusions and make recommendations.

Recognize legs of your three-legged stool. Design your presentation so that questions of this sort are preempted.

> *I was assigned to assess a department's performance relative to its business plan. I reviewed its performance by operating division, gathering data on various indicators for operating budgets, projects, capital spending, service delivery and customer service. For context and support, I interviewed key people. I prepared to present the results before the unit's management team attended by division managers.*

> *I realized that an objective evaluation can't be all sunshine and roses. There will always be shortfalls. Some divisions will perform better than others. This concerned me as I prepared my presentation. I knew that if the results didn't sit well with even one manager, how I arrived at the results could be questioned.*

> *Presentation day. I started by announcing that my presentation would come in two parts; the first explaining my methodology, and the second, the actual results.*

> *At the end of the first part, everyone agreed that the methodology was sound so I proceeded to discuss the results.*

> *The management team accepted the results, both good and bad, as well as my recommendations.*

> *After the meeting, one of the division managers approached me with a grin and said "I saw what you did."*

> *" What did I do?"*

> *" You got us to accept the methodology first so no one could question the results."*

> *I smiled. "Yes I did."*

> *He added, "It could have been a different story had you started with the results first."*

> *" Yes, it could."*

I realized early on that methodology was a key post in my three-legged stool. If somebody at the meeting questioned the methodology and cast doubt as to its validity, the subsequent results would likewise be questioned. By first taking care of the business of getting agreement on the methodology, I mitigated the risk of failure later on.

Now, suppose the management team didn't agree with the methodology? The only sensible recourse would have been to end the presentation midway and not present the results. I would thank them for their insights and suggestions and ask to come back and present another day.

Always ask yourself, "What success-critical premises have I made?" Make sure they're defensible and be prepared to defend them. Better yet, design your presentation in a way that mitigates the risk.

False messages

You must steer clear of the second, self-inflicted trap—that of getting caught knowingly delivering a false message. Some presenters get so cocky, they think their presentation skills can overcome bad information ("I can sell anybody anything"), so they do it anyway.

A school network engaged an advertising agency to conduct a weekend marketing event to attract recent high school graduates to enroll in one of their schools. The event was held at a large mall where the agency put up a show that featured young popular personalities. Song and dance numbers provided entertainment between marketing spiels about the school.

The agency was asked to present the results of the event to the board. Prior to the board presentation, an agency presenter and his boss had to present to the General Manager (GM) as a pre-board management review.

The presenter started by declaring the event a huge success, while flashing pictures of the event on the screen.

This claim unsettled the GM since the feedback from his own sources was markedly different. The GM preferred to let numbers speak for themselves and interrupted the presenter.

" Tell me, how many people attended the event?"

" About 8,000—give or take".

" Of the 8,000 who attended, how many do you think are recent high school graduates—your target demographic?"

" Ahhh, maybe 3,000."

"When you proposed a budget, how many recent high school graduates did you say you hoped to attract to the event?"

The presenter looks to his boss before answering, uneasy at where the questions were leading. He admits, "We targeted 30,000."

" So, you targeted 30,000, attracted 3,000 or just 10% of target while spending 100% of the budget. Would you say that's accurate?"

The presenter hesitated, then nodded.

The GM asks his final question. "So how can you say the event was a huge success?"

The presenter's boss hurriedly asked to end the meeting, saying they will revise their board presentation. They did. Their board report on the event changed from "huge success" to "lessons learned."

Ruthless? Perhaps. But in reality, the GM did the agency a favor. Those same questions could have been asked by a director at the board presentation, with disastrous results for the agency's reputation.

Don't ever knowingly present a lie and then rely on presentation or selling skills to get you through. There's bound to be somebody in the audience who will see through the fluff and point it out.

Board members in particular are often seasoned professionals. If you're lucky, they will be nice and not be as methodically ruthless at cutting down your claim, but they will know. They would see through the BS. Getting caught with a lie being packaged as truth is a career ender.

Just don't lie.

CHAPTER 12
Team presentations

Team presentations are fairly common, especially if you work in an organization that deals with large projects. The range of technical expertise needed to do a job may be so diverse, that the proposal presentation needed to win it can't be credibly delivered by a single person.

In such cases, bidding for work, or presenting project results, may be best handled by a team that can represent each of those areas of expertise.

The implied message when presenting as a group is that "we're better and stronger as a team." The team presentation has to reflect this message.

Easier said than done.

Presenting as a group has its own set of challenges. The goal is to impress and persuade—together.

Done right, your audience of decision makers will perceive a team of competent professionals that work well together. This perception will tend to positively affect their appraisal of the proposal itself—it's the halo effect all over again (Chapter 10).

Done poorly, you could end up with everything from inconsistent slides, misaligned messages and conflicting answers to questions—signs that scream the opposite of "team". If you can't get your act together in a team presentation, how can you be trusted with a large, complex project?

A team presentation delivered by n members has **(n+1)** more variables that need to be managed, which increases both difficulty and risk.

▶ Instead of just one person preparing his message, you now have n members contributing their expertise to the message, PLUS one person responsible for combining the messages into a cohesive whole.

▶ Instead of just one person designing his slides, you now have n members contributing slides, PLUS one person consolidating everything into a visually consistent, smooth-flowing presentation.

▶ Instead of just one person practicing his delivery, you now have n members practicing their assigned parts, PLUS the added task of practicing together delivering the complete presentation.

▶ Instead of just one person delivering his presentation, you now have n presenters, PLUS one who will act as the overall lead.

This reminds me of a rather frustrating experience working on a team presentation.

I was providing graphic support for a team that was given two weeks to prepare for a major bid. The five presenters—all very experienced—agreed early on who would take different sections. Each one had a week to compile their materials in advance and place them into slides. I was to gather and combine everyone's contributions and bring everything to brand compliance. I clearly stipulated that I needed at least two days to work on everything. That gave the team at least five days to practice and do edits before D-day.

Everyone complied except for one, who chose to ignore the schedule and didn't submit his material until the night before the presentation. Granted, he was one of the more experienced speakers in the group, it didn't sit well with the rest of the group that he didn't even show up for practices, with the excuse that he was too busy meeting clients.

He also assumed I would have the time to insert and work on his section at the last minute. In fact, when he finally called the night before and realized they were presenting the next morning, he said, "Oh, what the heck. I'll just wing it."

Just wing it? Seriously?

I imagine some of you might have had similar experiences.

So, how do you mitigate the risk and achieve your goal?

1. Limit the number of presenters (n) to the bare minimum. You may consider bringing non-presenting experts with you as support, but keep those to a bare minimum as well. Some clients are put off by an unnecessarily large contingent.

2. Assign a capable team leader who will be responsible for the entire presentation. Shared responsibility often results in unclear accountability. Team members carry their weight under the team leader's direction.

The leader:

▶ creates and orchestrates a plan, complete with objectives, roles, resources, timelines, milestones and deliverables.

▶ assembles the messages in collaboration with the team, ensuring that everyone understands the entire presentation—audience, purpose, messages.

▶ ensures that all the information needed to create the slides are provided with enough lead time to properly sequence them, do quality checks, edits, practice runs and further edits, should the need arise.

▶ often lands the role of lead presenter, delivering the introductory and closing remarks. The leader manages handoffs between presenters, keeps time, trimming individual presentations on-the-fly and stepping in if needed. The leader moderates the Q&A, referring questions to the right resource, and ensuring answers are correct and aligned with presentation messages.

3. Each member of the presentation team must know their part well. Moreover, everyone must understand how their part fits and contributes to the full picture. Everyone should stay attentive and supportive throughout the delivery, should someone in the audience unexpectedly interrupt with a question that would be best addressed by a different member of the team. If this happens, keep the response as succinct as possible so as not to derail the presentation's timing.

4. While each member is expected to know and practice their part, the team must practice the entire presentation together—repeatedly. Seeing the whole picture. Doing hand-offs. Giving feedback.

Keeping time. Simulating a Q&A with questions that are most likely to be asked. Including some unlikely ones. Have capable people outside of your team sit in on your practice runs who can give objective and unfiltered feedback then work out amongst yourselves how to best address any concerns.

5. Set aside enough time for preparation and practice. The more members there are in a team, the more lead time the team needs to prepare and be ready.

Sometimes, a presentation is best delivered by a team. Just be sure that the presenters prepared *as a team*.

CHAPTER 13

Practice, practice, practice

Let's assume you know your purpose and the messages you want to communicate to your audience. You've written the script, composed your slides, and proofread every single one of them. Your messages and your materials are ready. Now what?

There are five things you need to do before the actual presentation:

1. Know your presentation.
2. Time your presentation.
3. Practice your entire presentation.
4. Practice your entire presentation.
5. Practice your entire presentation.

I'm not being smart-alecky with the last two. I just can't emphasize enough that repeated practice is necessary. Oh, did I mention that repeated practice is necessary?

Know your presentation

It seems obvious but you'd be surprised how many people will go in front of an audience without being fully familiar with the presentation, its parts and its precise sequence. This typically happens when deadlines are tight and materials weren't readied in time. Maybe requests for the presentation were made at the very last minute. Maybe a boss had his staff prepare it but he didn't have time for an ample review. The presenter might have only seen printouts of the slides half an hour before getting on stage and still felt confident enough to just wing it.

These are all recipes for disaster.

You need to do the work and know your presentation inside out. Here's my three-way test for determining just how prepared you are:

Are you familiar enough with your presentation to know exactly what to say in each slide? Do you know the exact sequence of the slides? If the projector or computer were to break down or malfunction, would you still be able to continue, unfazed, without skipping a beat?

Do you know your presentation well enough to clearly explain the core message in under three minutes if you have to—or expound on it for two hours if asked?

Have you reasonably anticipated all the possible questions that might be thrown at you and are prepared with clear and direct answers?

If you can answer yes to all three, then you're ready. This amount of preparation will equip you with the confidence you'll need, and it will show in your performance.

Time your presentation

Before accepting a presentation assignment, one of the first things you want to know is how much time you will have for the actual presentation. This will largely dictate the coverage and amount of detail you'll be able to get into. As you gain more experience, you should be able to gauge, based on your speaking pace and mastery of the subject, if the time you're being given is realistic.

You need to realize that going over your allotted time is being disrespectful of your audience. You might find yourself being signaled to pick up the pace or proceed to wrap. The two most embarrassing things you could do (and I've witnessed this) is go on a rapid-fire blitz of your remaining slides in an attempt to make up for lost time, or do the unthinkable and beg for more time on the spot. These offenses should be punishable by law—considering the torture you would be inflicting on your hapless audience.

On the flip side, a powerful presentation delivered in less than the expected time is always appreciated. Remember—people don't attend these presentations for pleasure. If you can let them go early so they can catch up with other tasks or grab a cup of coffee— wonderful!

Time your entire presentation each time you practice. Try going through the entire thing in your mind, actually mouthing the words in a steady, deliberate pace that's easy to listen to and understand.

Recognize the difference between your talk time and your total time. Allot some time for a quick introduction, occasional ad libs and a few questions after. If a lengthy question and answer portion is expected, make the necessary adjustments.

Let me share a story about what could happen if you don't time your presentation.

> *Bill was an engineer tasked to provide the board with an important project update. It was a key item on the agenda and was given 30 minutes. He recounts his experience to a friend...*
>
> " *I prepared for three solid weeks and had 92 slides to go through.*"
>
> " *92? Isn't that a little too much for just 30 minutes?*"
>
> " *Well, there were a lot of points I needed to cover.*"
>
> " *So, how did it go?*"
>
> " *Well, I think it was around the 25-minute mark when the chairman interrupted me... 'Bill, how many slides did you prepare?'*"
>
> " *I go, 92.*"
>
> " *And how many more to go?*"
>
> " *I'm on slide 56 but I can go much faster.*"

" Before I knew it, my boss decides to cut in, summarizing the key points of my entire presentation then wrapped it up—all in less than 3 minutes."

Bill paused and added, "The chairman nodded afterwards and said 'Okay, we get it—thank you.' He didn't look pleased."

" So, you didn't get to show all your slides?"

" Unfortunately, not."

" I have one question. Did you practice and time your presentation? "

" I didn't actually practice, as in go through the entire thing."

" But why not?"

" I was too busy putting together the deck. 92 slides takes a lot of time to put together!"

It's obvious that Bill knew his material. He prepared—the best way he knew how, but unfortunately lost sight of the importance of time. He made two fundamental mistakes. He didn't identify and focus on the core message, and he didn't practice and time his presentation. He was lucky his boss was there to bail him out.

A few tips for controlling too much content

Realizing you have too many slides after investing a ton of time and energy preparing them, is painful to say the least. Spare yourself the agony by starting with a core message and expanding it incrementally as we've discussed in Chapter 3.

Still, if you ever catch yourself in the same predicament as Bill, with too many slides and too little time, here are a few tips you might want to try in a pinch.

1. Start by identifying the slides that deal with just your main discussion points—stuff you simply can't leave out.

    For the main slides, ask yourself, do you have multiple slides hammering in the same message? Can you be more succinct and use fewer words? Are you over-illustrating a point?

    These questions allow you to simplify and trim down your main slides further. Now try to simulate and time your presentation using just these trimmed-down main slides.

    If you're still over the time limit, then it's probably time to start cutting down on your messages—and your slides. Review your core message and focus on getting that across. Carefully weigh which slides you can do without, and which ones are key. Cutting down on supporting messages should significantly prune your talk time. Continue pruning until a practice run clocks in within your allotted limit. When you've reached that point, cut just a couple more. This will give you a bit more breathing room.

    Throwing out slides is a painful but necessary exercise—one that you will learn to avoid with experience. You have to be ruthless. Don't fall in love with some of your slides just because you worked particularly hard on them or they have really cool graphics.

2. You don't have to actually delete the remaining slides. These can still prove useful when a question is asked, and the answer is in your slides about supporting messages and details that you've cut out. Just use PowerPoint's **Hide** function so that these slides are still in your presentation file but won't be shown as you play your presentation.

   Now, if anything on those hidden slides ever come up, say in the Q&A, it's a simple enough thing to hit the **ESC** key, quickly scroll to that slide, and hit **Play**. Sure, it isn't the most elegant solution, but it allows you a better chance of covering the most important points. Leave it to your audience to ask about other details later if they feel it's important. Don't preempt the need and put yourself in jeopardy with time.

3. If you have advanced PowerPoint skills (or know someone who does), you can program unobtrusive "hot buttons" on a slide that will allow you to jump to other related "detail" slides as an option. Just make sure you program Return buttons on those slides so you can quickly jump back to the spot you jumped from.

Food for thought

Remember, in Chapter 19 we said it doesn't make sense to set a cap on the number of slides. The assumption here is that you've trimmed your messages to just the right amount of detail for the allotted time. So, if you find yourself with more slides than you can go through within a given time during practice, the question you should be asking yourself is "Do I have too many messages?" or "Did I go into too much detail than what's called for?"

Practice your presentation

Have you ever presented poorly then started banging your head on the wall for the rest of the day thinking "If only I could do it over again!"

That's why practice is SO important. Practice will force you to go through every point, identify any parts that you might be struggling with, and find ways to get through them. I guarantee you will improve with every pass until you reach a point where it's smooth and confidence takes over. This will translate into an effortless overall delivery.

Is there a method to practicing? Here's a suggestion.

Start by practicing in sections. Practice your introduction. When you've mastered that, practice presenting the body. You may have to break the body up further, probably moving from describing a problem to your analysis on to a solution. Then practice your closing and going for your win.

Do a pass where you stitch all the sections together. If you hit a bump, take note, then do it again. And again. And again. Start by practicing the words in your head first. Then practice by actually voicing the words.

Once you've stitched all the sections together and voiced everything, timing each pass to make sure you're within limits, it's time to call in a friend or colleague for honest feedback. You can also take a video with your phone and take notes on what needs improvement.

You can also do as I do—visualize your first few practice sessions in your head. I narrated my first experience with visualization as a practice tool in Chapter 14. Read on to the next section on Visualized Practice for more on this technique. It might surprise you just how effective it is.

How long should you practice? Let me rephrase that. How long do master presenters practice?

Steve Jobs was widely considered as the master presenter. He made exciting Apple product launches look effortless. He was not only possessed of natural talent but he also knew his products better than just about anybody. He would get deeply involved in crafting the messages in a launch, down to the images and a few well-chosen words. His message was crystal-clear in his head and he designed how he wanted his audience to react at every single juncture. Nothing was left to chance.

One would think that with all this talent, knowledge and resources at his disposal, he would need little to no practice.

And yet he did…with an almost maniacal obsession.

In Carmine Gallo's book *Presentation Secrets of Steve Jobs*, he writes:

> " *Jobs can make solid eye contact with his listeners because he practices his presentations for weeks ahead of time. He knows exactly what's on each slide and what he's going to say when the slide appears. The more Jobs rehearses, the more he has internalized the content, and the easier it is for him to connect with his listeners."*

Even for the master, no less than weeks of practice would suffice.

If you think you can get away without practice and still do a great job, think again.

Photo by Acaben, cropped by Kyro - Flickr, CC BY-SA 2.0

Visualized practice

Nothing beats good-old-fashioned practice but is there another way that's equally effective and easier?

Australian psychologist Alan Richardson conducted an experiment on sports visualization with three groups of students chosen randomly. On day 1, each student shot 100 basketball free throws, which became their individual baseline.

For the next 20 days, Group 1 practiced shooting free throws for 20 minutes. Group 2 did not practice and was, in fact, instructed to not think about basketball. Group 3 only visualized practicing shooting free throws for 20 minutes a day. Sometimes they "missed" their shots. Sometimes they didn't. But they "practiced" in their heads.

On the last day, each student shot 100 free throws and the results compared with their Day 1 baseline. The results were surprising.

It would seem that visualized practice was almost as effective as physical practice!

I can't speak with authority on visualized basketball practice but I've always known visualized presentation practice to be very effective. I count two hours of visualized presentation practice as equivalent to an hour of actual practice. I like to do a mix of both, everyday, leading up to D-day.

Nights before D-day, I would visualize my presentations just as I'm about to doze off for the night—instead of counting sheep. Who's to say my mind didn't continue practicing until well into morning?

Visualized practice works best when you've crafted your messages well, composed your slides and timed your presentation. You have just the right collection of stories and planned your pauses. You've done the work and now need only deliver your best presentation ever.

That's when you start visualized practice.

In the next chapter, you'll read the story of how, as a college sophomore, I stumbled on using visualized practice for presentations. My speech came out exactly as I had visualized it. That also marked the day that I overcame my terrible fear of public speaking.

Group	Basketball foul shot practice instructions	Improvement
1	**Actual Practice** *20 mins, 5 days/week for four weeks*	**24%**
2	**No Practice** *Don't even think about basketball*	**None**
3	**Visualized Practice** *20 mins, 5 days/week for four weeks*	**23%**

" **I've always believed that if you put in the work, the results will come. I don't do things half-heartedly. Because I know if I do, then I can expect half-hearted results.**"

Michael Jordan

CHAPTER 14
Stage fright

It's been said that many people fear public speaking more than death itself.

In high school, I was terrified of speaking in class. I remember trying (and failing miserably) to make myself invisible and not catch the teacher's eye, hoping that I don't get called to recite or answer a question. I often knew the answers—I just dreaded speaking in class.

But one event in college changed all that.

As a college sophomore, I enrolled in a required Speech class. Students took turns speaking on stage in front of a class of about twenty.

My first few 3-minute talks were awful. I fiddled with a pen while talking. I tried to correct this by crossing my hands. Then I tried tucking my hands in my pockets. I stammered. I said "ahhh" every other sentence. I was so nervous I often forgot what to say next. I would mumble my words. I spoke softly. When I did speak, I spoke very fast, as I usually do when stressed.

Fortunately, those were just practice runs. My course grade depended mainly on a final 15-minute speech to be delivered before the end of the term.

I decided to put all my free time preparing for that final speech. I had to do well as keeping my scholarship depended on it.

I chose "The Threat of Extinction of the Blue Whale" as my topic—something I was passionate about. I did my research. I created my visual aids on large sheets of Manila paper (PowerPoint didn't exist then) and wrote my speech. Everything was ready two weeks before my scheduled speech. Everything but me. I didn't feel ready.

A week later, I already had my speech memorized and timed. I was still terrified and I had just one more week to get ready.

Then I started to really practice. I spent hours seated under a giant acacia tree on campus, on a small hill overlooking a soccer field. With no one within hearing range, I spoke the words. I heard myself going through the speech. My words were monotonic at first. Later I slowly added deliberate variations in volume, tone, pitch, pace and emphasis. My passion for the topic started to show.

Next, I started to visualize myself delivering the presentation. I would close my eyes and practice. In my mind, I could see myself on stage, speaking confidently while explaining the visual aids. Each visualization was part practice, part wishful thinking. It was like watching a movie in my head many times and I was the star. I was great. I could do no wrong.

The day came. It was my turn to speak.

In truth, I barely remember the actual presentation. It was all a blur—I must have zoned out of body while presenting or maybe I was possessed.

I do remember realizing, as if waking up, that I was standing on stage in front of a class that was clapping. Clapping! I also remember the teacher saying "That was an excellent speech."

As I took down my visual aids, I wondered "What just happened? Why did they clap?"

Everything apparently went exactly as I had visualized it. Got an A- in that course.

That day, I overcame my fear of public speaking.

I learned that I just needed three things to overcome my fear. I must have something of value to say—something I'm **passionate** about. I must be **prepared**. I must **practice**.

The more important the presentation, the more preparation and practice I needed to do—from knowing the content, to hearing myself speak the words, to visualizing the final outcome.

No one taught me this. I just stumbled on this formula out of desperation.

How can preparation and practice overcome stage fright? I believe that the fear springs from the possibility of doing poorly and getting embarrassed or ridiculed in public. It's all in our head—this fear of messing up.

That's exactly what practice addresses. You get to mess up big time in a safe environment, with only you knowing that you messed up and you get to do it over and over again.

The more you practice, the less mistakes you make and the less potential to mess up.

If you practiced and visualized enough, you would be walking to the stage saying to yourself "I've done this hundreds of times. I got this." The chances of messing up and getting publicly embarrassed decreases with practice, and along with this, your stage fright.

Does anyone ever really overcome the fear of public speaking?

Maybe not. Even now, I still feel my heart pounding every time my turn to speak nears. Only now, I think I'm more excited than afraid. Or maybe the fear is there but I've learned to overcome it. Either way, it gets my blood pumping and gives me a rush. I'm ready and eager to go on stage.

Passion for the message. Preparation. Practice.

Works for me.

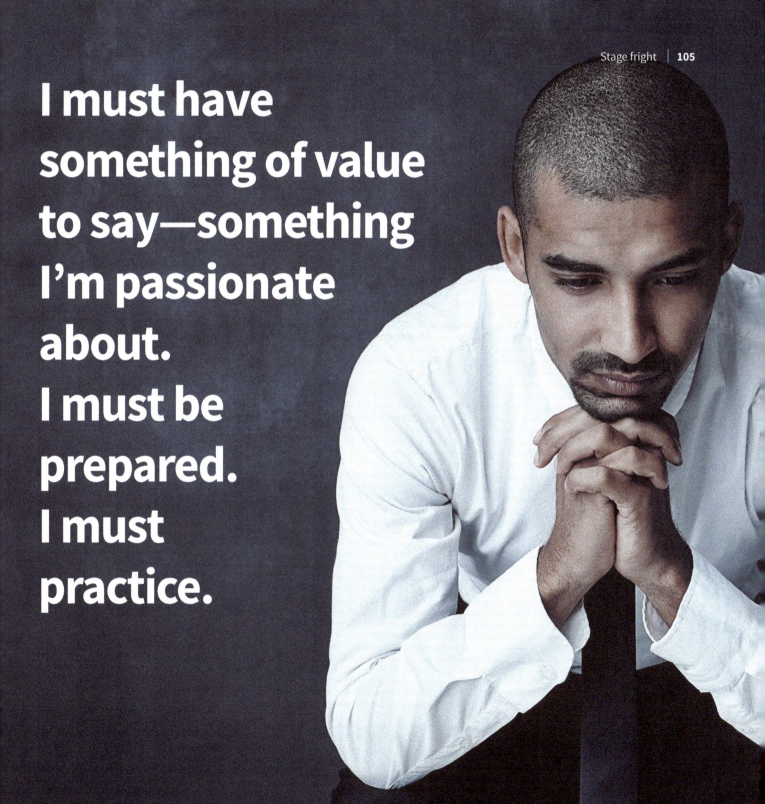

I must have something of value to say—something I'm passionate about.
I must be prepared.
I must practice.

CHAPTER 15
Your first two minutes

Let's take a look what happens in the first two minutes of your presentation.

Do first impressions really matter?

I still remember my very first job interview in the '80s. A large multinational company was recruiting for wireline field engineer trainee positions. They posted an open invite for college seniors to come to a five-star hotel for an orientation followed by job interviews.

Many came, mostly seniors like me still months away from graduation. I was waiting in a hallway for my turn when the interview room door opens. An applicant was being let out of the room. He was upset and was about to argue with the recruiter, when the recruiter angrily said, "if you don't know how a full bridge rectifier works, you don't know anything!"

Visibly irked, the recruiter then looks our way and says, "Next!"

Scary. The guy ahead of me stands up. He was slightly bowed as he walked, probably intimidated, his left hand rubbing the back of his neck, while he tentatively extends the other hand towards the recruiter.

The recruiter shakes his hand and says, "Thank you for coming. Next!"

The poor guy didn't even get inside the room! It seemed that in one quick second, the recruiter sized him up and found him to be not worth the time interviewing.

Brutal.

I was next. As I approached the recruiter, his face lights up. "I remember you. You were seated in the front row during the orientation earlier. You were smiling the whole time. Come on in."

First impressions …. What worked well for me worked very badly against another. I aced that interview and later got offered the job, but that's another story.

It was a life lesson learned early—first impressions do matter. And a smile? That works too.

Now, let's assume you're presenting before an audience who doesn't really know you.

Let's focus on the first few seconds before your talk, starting from when you're called to come on stage or in front of the board of directors.

As you're walking up and all eyes are on you, your messaging starts immediately, and it's all nonverbal. In those first few moments, your audience starts mentally sizing you up, creating their own personal impressions. Do you look credible? Interesting? Trustworthy? Or is this all a waste of my time?

These first impressions set the tone for the rest of your presentation, so you want to make it positive. Here are a few **Do's**:

Do look your best, which means dressing slightly more formally than the audience. When you get to be as famous and admired as Steve Jobs, you can rock the blue jeans and black mock turtleneck. Until then, dress to impress. Good dressers make an audience feel respected and important. Some count it as a simple but expected courtesy.

Do stand and walk tall. Straighten up and pull your shoulders back, slightly. You'll come across as proactive and confident. Slouching makes you look timid and unsure. Go to a busy public place and observe the posture of people you don't know. What are YOUR first impressions of people who 'walk tall' and of people who slouch?

Do smile. A warm smile shows confidence. It implies you're happy to be there. It helps build an initial rapport with your audience and makes you look likable.

Do establish eye contact early. Good eye contact suggests you're open and can be trusted. Try sweeping your eye across the room from time to time so no one feels excluded.

Here are a few **Don'ts**.

Don't cross your arms. Don't place your hands on your waist or in your pockets.

Don't start with a microphone test—unless irritating your audience is part of your plan. "Can the people in the back hear me? Yes? No? Testing…testing…1-2-3…" This sound check should have been done before there were people in the room.

Never start with an apology. "I'm sorry but the PowerPoint file I intended to use keeps crashing, so I'm using an older version." Some things, people in the audience don't really need to know.

I know it seems like a lot to remember for the first sixty seconds and you may not remember everything when the time comes. But the point is to be aware that these things do count and is all part of the journey of becoming a great presenter.

Here's a thought. If you have a smartphone, try recording yourself in your first couple of minutes, starting from before you come on stage. Then check through the Do's and Don'ts. You won't know what you don't see and you can't fix what you don't know.

Your opening

You've walked in and made your first impressions. You are now standing in front of your audience and ready to start. What's next?

Start your presentation by answering 'questions' that are most likely forming in your audience's minds.

They expect to know who you are, what you'll be talking about and what qualifies you to talk about it. This establishes your credibility as a source of information or opinion.

They will likely want to know how long the talk will take and why it's important for them to listen. What can you do for them? Unless they see the value, why should they even give you their time and attention?

You will want them to know what the purpose of your talk is and what you will want from them when you finish. This shows that both the audience and you have something to get out of the presentation—a win-win.

You might also want to state your preference on audience questions. Should they feel free to interrupt you with questions anytime? Should they hold off until Q&A after? Read Chapter 19 for my suggestions regarding answering audience questions.

Try to write out your first few lines so that you're comfortable and off to a good start. Be clear and concise. You should be able to open your talk and answer these first few questions in your audience's mind in less than two minutes.

" **Everyone smiles in the same language** "

Unknown

Victor Panlilio Photography

CHAPTER 16

Let's have a conversation

Audience engagement is key to a successful presentation. An engaged audience listens, shows interest, processes what they hear and reacts in ways that provide you with feedback. Can you tell when an audience is engaged?

If you've been presenting for as long as I have, you start to observe consistent clues.

One is when your audience don't break eye contact, don't allow themselves to get distracted by text messages, or don't keep checking their watches. In fact, a really engaged audience will lose track of time, and will want to keep you there beyond your allotted time.

Some clues are a little more subtle—an unconscious nod or a smile of tacit approval right after you say something they agree with. The act of scribbling notes on a pad after a particularly interesting statement, or, my favorite; a wide-eyed response showing enthusiasm or excitement upon hearing or seeing something particularly interesting, often accompanied by a faint smile and subtle shift in their sitting position.

Towards the end of your talk, an engaged audience will ask relevant questions and you get the sense that they're not just asking to test your knowledge, but that they genuinely want to know the answer. It's clear evidence they were listening.

You've probably heard this many times—that communication is a two-way street. If you think about it,

everything I've just described is just like a conversation. I know that sounds contradictory to how we've characterized presentations in the earlier chapters—that it's a one-way street and you're the one in control. But hear me out.

Unlike regular conversations, like the ones you have with your officemates during a break, your side of the "conversation" is verbal, while the other side, largely non-verbal. Sound silly? Not really. Entire non-verbal conversations take place between people every day—just ask any couple who's been together for more than twenty years. Suddenly, the prospect of a verbal/non-verbal conversation is not so strange after all.

When you converse, the other person listens and responds. You get feedback and you respond. Through this give-and-take, messages are sent and received. There's full attention and active presence on both sides. No one falls asleep on you in a conversation. At least you hope not.

This is exactly what we'd like to happen in a presentation, and here's how it works.

As the presenter, act and speak as if you were conversing with the audience. If your audience senses the conversational dynamic, they will naturally respond, but mostly in nonverbal ways. You'll be surprised at how interactive it can get—if you just pay attention. When properly engaged, people will respond intellectually, think about what you just said and connect with your stories emotionally. Your audience will lock their gaze, pay attention, listen, maybe squirm, frown, smile or laugh. That's their conversational response! Your goal should be to sustain that whole dynamic throughout your presentation.

Now let's break it down. What are the actual mechanics of turning a presentation into a conversation? You can start by imagining the things you would normally do in a conversation as well as the things you wouldn't.

Eye contact

Imagine conversing with a person who doesn't look you in the eye. Clearly, it speaks to that person's level of interest, trust, openness and willingness to engage.

To a presenter, eye contact isn't optional—it's an absolute must. When presenting, move your eyes from one person to the next, scanning the room. Don't leave anyone out. Establish eye contact from the moment you take your position, and maintain it throughout your talk.

For some, this comes naturally. For others, it may take quite a bit of practice before becoming comfortable with it. But there are no two ways about it. This should be at the top your list of things to do and learn. Remember—you can't have a conversation without eye contact.

Reading isn't presenting—so don't read your slides!

Some presenters turn to the screen and start reading from their slides, word-for-word. This terrible habit reduces a presentation to the rough equivalent of an elementary school reading exercise… "Okay, class! Let's all read together what's on the board…"

Let me tell you a story.

I arranged for a one-on-one presentation between a retired professor and the chairman of an education agency. The professor had substantial experience developing online training courses and had a proposal for the agency.

I brought the professor to the chairman's office and introduced him to the chairman. The professor placed his laptop on a meeting table and launched his slide deck as the chairman looked on.

I was surprised when the professor went through the first few slides by reading the bulleted sentences almost word-for-word.

By the fourth slide, the chairman interrupted him.

" Wait", he said. "I can read faster than you."

The chairman took over the laptop and quickly clicked through the slides while we quietly watched. He was quite the fast reader.

A minute later he says, "Okay, I get it. The two of you can discuss how to move this forward."

End of meeting.

The chairman didn't mean to be rude. He was obviously a very busy man and felt he could do things more efficiently. Why listen to someone read through a presentation word for word for fifteen minutes when he could read it in under one minute? It would have been a waste of his time.

Moral of the story? If all you're going to do is read your slides, then save everybody some time. Just give them a copy of the deck, give them five minutes to read through it and then ask if they have questions. At least, you've provided *some* kind of value.

The whole practice of reading off a screen is lamentable and will reflect poorly on you. It smacks of laziness, lack of skill, preparation and overall professionalism. If you have to read off a screen like you would a script, then people will start to question in their minds if you really know the material well enough to speak about the subject unrehearsed. Did someone else write this material for you? Did you even prepare for this talk?

Never turn your back

An added negative side effect of reading your slides word-for-word is that you invariably turn your back to your audience. Every time you turn away, you break eye contact and are essentially surrendering control.

If you've prepared your slides yourself and really practiced, you shouldn't even have to turn away—just advance to the next slide while speaking, without skipping a beat.

If you're using the slides to cue you during your talk and don't have a second monitor facing you, you can take a quick glance at the screen, but the operative word here is 'quick'—almost like a blink. Eye contact isn't broken by an occasional blink.

Should your audience read or listen?

We've established that YOU shouldn't be reading your slides. How about your audience? Should THEY be reading your slides?

Your audience should be listening to you and what you have to say, instead of reading. Your slides are there for support—to enhance what you're saying, not take its place. When they do read your slide, it has to be a very quick read, maybe half-a-second or so, after which they revert back to listening to you.

What does half-a-second of "reading" mean?

That translates either to one picture, one word, or a short phrase or bullet. A complete sentence will begin to compete for your audience's attention—do they read or do they listen?

A page full of sentences and bullets? No contest. It might take your audience 20-30 seconds to read the entire slide. They will read ahead of you. During this time, they are NOT listening to you.

Your strategy, therefore, is to flash on your slide only information that is immediately relevant to what you're saying at that exact moment.

In the audiovisual world, this is referred to as "audio/video lock." What you see has to be in perfect sync with what you hear. When you lose the lock, it throws off the audience. This is why it's so important to not only control the timing and release of information one bite at a time, but make sure each bite is chewably small. A word or phrase will do, as opposed to complete sentences. This way, your audience doesn't spend more than a split-second reading, then they can continue focusing on the more important task of listening to you speak.

More than words

Messages aren't just words. The way words are delivered is just as important as the words themselves and makes a material difference in how the message is understood. Were the words whispered? Were they shouted out? Did you get a sense of seriousness, urgency, empathy? Or was it lighthearted and upbeat? Delivery is key.

Perhaps the worst example of poor delivery is having to listen to someone speak in monotones, with an extra helping of reading just about everything off the screen. Just shoot me and get it over with.

We rarely, if at all, carry on a conversation in monotones. Depending on what the subject of conversation is, we instinctively change the way we speak because, well, it's just natural. And it comes naturally because we want the person we're speaking with to know how we feel about what we're saying. So why should it be any different when we're delivering a presentation?

Imagine a monotonic delivery of a presentation before a large audience by a senior manager reading off a prepared speech that he probably didn't write himself. Maybe his staff wrote the speech and prepared the slides—the same staff clicking through the slides in sync with his reading. Fully intending to just read his presentation, the manager doesn't really prepare and doesn't "feel" the words, so that the delivery comes out as monotonic. Many times, he doesn't even look up.

He's not conversing with you. There really is no you—just him and the words of his speech. He's going through the chore of reading a speech … and you, the agony of having to listen to his monotonic delivery.

I've encountered this lose-lose scenario many times. Terrible presentations.

Having a conversation is not just about words. The message is in the words *and* your voice—your pitch, tone, volume and pace. The message is also in your body language—your arm gestures, facial expressions, eye contact, the way you stand, the way you move.

All these come out naturally in a conversation, *even without practice*. Think how much more effective you will be when you do practice.

Level up on your presentations. Engage your audience in a conversation.

CHAPTER 17

Power tools

You are now in a conversation with your audience. They are engaged. Here are some simple, everyday things you can do to raise the level of engagement up a couple of notches. These techniques are derived from certain instinctive human behaviors and are designed to pull your audience in without them knowing it. We'll call these your power tools.

Ask a question

When you're thrown a question, either directly or in a group, your brain involuntarily goes on autopilot and starts rummaging around for an answer. You just can't help it. It may be tied to an innate, often secret desire to prove ourselves smarter than the next guy. This will go on for a fraction of a second until you decide you don't know the answer, or simply don't care.

So, how can we leverage this impulse?

You can ask your audience a question that you'll eventually answer yourself. Or provoke interest by asking something rhetorical to get people thinking. Use the question to pique interest, in effect, pulling them in by polling the group for answers. This device works all the time.

For example, as a lead-up to a discussion into the causes of a certain problem, you can pose the question, "What do you think are the causes of this problem?"

This grabs your audience's attention back to your topic, by getting them to search their personal knowledge base, "Do I know the causes of this problem?"

This search happens in an instant and might not even have a definite result but the search happens nonetheless. That momentary engagement is a way of re-centering any minds in the audience that may have wandered in some other direction.

Further into the presentation, you can ask, "How do you propose we solve this problem?" This will trigger yet another round of engagement and interaction. Keep your audience synchronized with where the presentation is headed.

For greater effect, you might try pausing for a second or two before popping the question. The short silence will intensify attentiveness, especially if the pause is somewhat unexpected. Giving your audience a few seconds after asking a question will give them time to think.

Now before the silence gets uncomfortably long due to people still figuring out the answer, step in by either rewording the question or dropping a few hints here and there. Don't let the silence stretch out any longer than 5 seconds tops, otherwise, your audience will start to feel ignorant or inadequate.

Pause

We all naturally pause in conversations. It's a chance for the other person to react, think or respond. It's essential to making conversations two-way. It's also a way of showing courtesy.

When speaking before an audience, use well-placed pauses to control the pace of the presentation and help bring out a more natural, conversational quality.

So, when and how do we pause?

The very first pause in your presentation should be at the very start, before uttering your first word. Pause. Let the audience settle down and focus on you for a few seconds. Take these few seconds to establish initial eye contact. Let your eyes slowly scan the room. An intentional pause that, in your mind, says, "I see you. Can we talk?"

It's a sneaky way of communicating without saying a word.

I've never been one to kick-off with a loud voice to signal an unsettled group that the presentation is starting. Maybe it's the professor in me. I've always preferred establishing eye contact with my students first then wait 'til everyone has quieted down and all eyes are on me, ready to listen.

Then I start.

There are three other ways pauses can be used to good effect.

Short pauses, like commas in a sentence, can give your listeners a brief time to process what you just said. These become more necessary the more complex the subject matter gets. Be sensitive to your audience's facial expressions and body language. It's easy enough to see when they look a little puzzled, unsure or overwhelmed with what you just put forth—be it something you said or a slide you just flashed. Read the room. Don't just keep plugging away, oblivious of what your audience is communicating to you non-verbally, for fear you'll run out of time. Pauses are a good way of helping your audience catch up, especially if you see them taking down notes. Pause too long, and the audience might interpret it as an invitation to ask a question to fill the void.

This short pause can also come after a rhetorical question. It gives your audience time to ruminate or chew on a truth they otherwise might not have arrived at by themselves. Make room for possible audience epiphanies—quiet 'aha!' moments that need a little more time to let something poignant or profound, sink in.

Another kind of pause I use is slightly longer, to help break up a long presentation into sections. Like breaking up a book into chapters.

You can for example, end the part where you explain the problem with a 3-second pause, as if in deep thought, then proceed with "Now that we've clarified the problem, let me present the solution." Another pause.

You've ended the problem discussion and segued into the solution discussion. This allows your audience to set aside the "problem box" and get ready to listen to the "solution box."

Or maybe you explained at the start that the talk will be in three parts. At the end of the second part, you can say, "That ends Part 2. Let's move onto the final part, Part 3." Then pause.

A third type of pause is much longer, which some might call a pregnant pause. This happens right after I end the presentation with a request for approval of a proposal. This type of pause is a way of handing over the ball to your audience. Consider using this before a small group of decision makers. I don't recommend this in presentations before large audiences—the void will immediately get filled-in with random chatter.

The extended pause will trigger someone, perhaps the chairman who's present, to chime in with an appropriate response. The book's final story in the Epilogue shows such a scenario.

Tell a story

Aside from trying to answer questions, another strong instinct people are hardwired for is connecting with stories.

You see it from childhood, kids gravitating to stories before bedtime—or pretty much any time. Holy books use parables to teach life lessons. Before written history, oral traditions through stories were passed on from one generation to the next. Nowadays, people will pay good money to watch a movie and while rumors and gossip are a huge waste of time, many indulge in them just the same—because a good story is hard to resist. Like it or not, we are all, without exception, story junkies.

When people listen to your presentation, chances are, they won't remember the litany of terms and numbers you presented. They might not remember your name. They will, however, remember a good story. If it's accompanied by an emotional component that your listeners can relate to, even better. Emotions make people remember things more effectively.

There is compelling scientific evidence that the human brain, one part in particular—the amygdala, considered to be the seat of human emotions—is stimulated when we listen to a good story. Interesting character-driven stories in particular, cause the secretion of oxytocin into the bloodstream, turning unsuspecting listeners into emotionally-charged beings who are more likely to take action and make decisions. I recall an extreme case when, after watching Rocky for the first time, I just felt compelled to go and work out for no apparent reason. But the overarching point to all this is the fact that information received while accompanied by an emotion, tends to be remembered the most.

This is powerful stuff. So, how can we use this information to deliver compelling presentations?

Your stories don't have to be particularly heart-wrenching to be remembered. Emotions come in different flavors and varying degrees. You can start simply by introducing a topic with an interesting story. Let's say you have a talk about inventions.

Some of the greatest inventions in the world came about as a result of an accident. The key is recognizing what the accident is telling you, and go from there.

While conducting an experiment, an engineer from Canon absent-mindedly rested his hot soldering iron on the needle of a syringe filled with ink. He noticed that fluid was being ejected from the needle. He was able to recreate this accident, which led to the invention of a new type of inkjet printer, the bubble jet.

Here's another one.

> *Nitrous oxide owes its dubious moniker, 'laughing gas,' from the days when it was a party toy where people would inhale enough of it to induce uncontrollable howls of laughter—for fun. But a friend of a dentist took too much of it and gashed his leg but didn't even notice until much later. The dentist realized that this 'party toy' could be used in his profession and thus was born one of the earliest forms of modern anesthesia.*

The audience might not remember the rest of your talk when they get home but they will remember your stories. They will retell it with "Do you know how modern anesthesia was invented? People used to inhale laughing gas in parties..."

You can also tell your own story, a personal experience to establish yourself as an expert.

> *People from the smartphone generation just don't realize the remarkable strides the electronics industry has achieved in the field of miniaturization.*
>
> *In the early 80's, I was responsible for maintaining an early-model CT scanner. It had one megabyte of image processing memory spread across a dozen electronic memory boards. Each board cost $4,000—about $12,500 today. The entire memory was housed in a huge metal box the size of a package air conditioner.*
>
> *Nowadays, you can buy a 128-gigabyte flash drive, roughly 128,000 times more capacity, in a package the size of your finger, and for the price of a fancy dinner.*

Doctor and Mrs. Syntax, with a party of friends, experimenting with laughing gas. Source: © Wellcome Images.

I remember giving a class lecture on transducers to electrical engineering juniors. I started to lose the class when I talked about ultrasonic transducers and electric pulses. *"Electrical pulses applied on the transducer produce ultrasonic waves…"* I lost them, not because they're slow, but because they just lost interest. If the brain is wired to pay attention to exciting, emotional stuff, it's equally hardwired to ignore anything boring.

I changed tack.

> " *You all know that ultrasound machines are used to take a peek at unborn babies, right? Here's how it works."*

> *I started drawing an outline of a very pregnant woman on the blackboard and then an outline of a baby inside the womb. I also drew a transducer close to the baby.*

Almost immediately, the entire class sat up straight and leaned forward. They looked amused at the drawing and interested at what would come next.

> " *An electric pulse applied to the ultrasonic transducer produces ultrasonic waves."*

> *I drew waves coming out of the transducer, hitting the baby and then drawing the signals that get reflected back.*

> " *Since the baby's different organs have different densities and depths, the reflected signals change in varying degrees and are read back into the transducer. The transducer converts the changes in the original signal back into electrical signals. These signals are processed by a computer into an image of the baby."*

> *I could see that the class understood and that they enjoyed learning something new.*

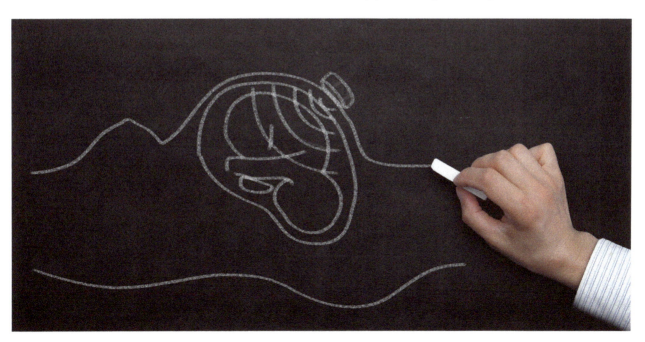

Seeing your students sit up, lean forward, listen intently and enjoy learning is one of the joys of teaching.

Sprinkle your presentation with stories to connect with your audience on a more emotional level. Tell your story naturally, informally. They will listen and be more attentive. More importantly, they will remember.

Live the story

Storytelling is a powerful tool. You can turn it up a notch by having your audience experience the story themselves. You let them be part of the story and live the story through a dramatization or a reenactment.

> *Bill Gates talked about the global challenges of eradicating malaria in a 2009 TED Talk. He had maps that showed malaria being eradicated in rich countries but was still a problem in poorer ones.*

He explains, "Now malaria is of course transmitted by mosquitoes. I brought some here, just so you could experience this …"

He opens a jar of mosquitoes and let them loose. You could see a mix of shocked and smiling faces in the audience, a brief "what does he think he's doing?" moment.

" We'll let those roam about in the auditorium a little bit."

The audience laughs. He continues, "There's no reason only poor people should have the experience."

The audience laughs, claps their hands.

He then assures the audience, "Those mosquitoes are not infected." and continues with his talk.

The audience was laughing. Amused. Relieved maybe. Definitely engaged.

I have another story—a personal one.

I was invited to speak about my experience migrating to Canada. Part of my talk was how I struggled with finding my first job.

> *" For more than three months, I applied online for thirty-three job openings, submitting about two to three a week. I felt qualified for each one. I customized each application. Each one, an honest and serious effort. Still, I didn't get any bites. No email. No calls. Nothing. I started to worry. Money was running low. I figured that if I didn't get a job offer in the next four weeks, I would have to leave Canada and go back home. Several times a day, I checked my email. All day, I would wish for the phone to ring."*

<short pause>

" I waited for the phone to ring. I waited ... and waited..."

< dramatic pause, I breathe in and let out a sigh ... >

" Then one day…"

A PHONE RINGS!! The sound fills the room. It caught the audience by surprise. Where is this ringing coming from? Did someone leave his ringer on?

I tell the audience "Excuse me", take out my cell phone then acted as if I was answering the call.

" Hello?"

A woman's voice comes over the sound system, " Hello. May I speak with Antonio please?"

The audience realizes what was happening. They were listening in on a private phone conversation.

" Speaking. May I know who's calling please?"

" Hi, this is Lisa with the City of Calgary. I'm calling about your application for the Project Specialist opening at city hall."

I act like I was screaming silently and throw a couple of fist pumps to the air.

The audience smiled and laughed, visibly amused. I had shared the anxiety of finding my first job as a new immigrant and of endless waiting for the phone to ring. Then I let them into my world through a reenactment—that happy moment when I got the call for my first job interview. (I got that job, by the way.)

People approached me after my presentation, still amused. "That phone call was a nice touch." "That was totally unexpected. I was looking around to see where the ringing was coming from. Who did the voice of Lisa?" I told them how I arranged for my daughter to act as the caller. I gave her a short script. She had a microphone with her when I started my presentation. She also had a ringtone ready to trigger when she hears me say "Then one day …" That was her cue.

I hope they remembered other parts of my talk, but that phone call was what really stood out.

Telling a story engages your audience. Letting them experience it with you through a dramatization, more so if it's a personal one, is even more powerful. It's unique, it's entertaining, but most of all, it's memorable.

Show-and-tell

Sometimes, even the best presentations need that little extra boost to make them even more convincing. Show-and-tell is one tool you might want to try.

I once argued that the supposedly secure ID card of an agency I was with could be easily faked. This was strongly challenged by the department producing the IDs. I was asked to present the issue before the board.

After presenting technical research supporting my statement, I erased all doubts when I dumped two boxes full of fake agency IDs before the board chairman.

I passed out the fake badges to other board members. While they held and examined them, I explained that these fakes were confiscated by another agency. It was physical evidence from a credible source.

I added, "There are many more where those came from."

Some of the IDs were bad replicas but some were astonishingly good. And there were two boxes of them.

There were no more questions. The board was swayed.

Research, physical proof that the board held in their hands and a touch of drama. The question was answered and was never raised again.

How about live demos?

Some try to do a show-and-tell using an actual demonstration of a product or application. Unless it's actually requested, I've always found this to be risky and will only attempt it after I've mitigated as much risk as I can actually control.

What are the risks?

First off, you have to be very familiar with the device or system you are doing a live demo on. Live demos have a way of coming up with something unexpected. The application might hang, or a connection gets lost, or a moving part just freezes on you. This is especially true with live demos of prototypes. If the unexpected surprises even you and you don't quickly manage it, or worse, if you get visibly rattled, then your live demo fails.

Never do a live demo on something you're not thoroughly familiar with, and not before you've successfully gone through the entire demo many times, and have fully anticipated Murphy to be lurking around.

If you're doing a demo of something like a desktop software application, for instance, be aware that user interfaces (fonts, graphics, on-screen buttons) are designed to be read on a screen that's a foot or two away from the user's eyes. Unless you're doing a demo up-close for just two or three people, your audience will likely be straining and squinting. Others may not even be able to get close enough to see the demo, even if they squint.

If you're running the demo on a system other than the actual one you've practiced on, you always run the risk of missing some critical setup detail that can cause it to misfire. Missing fonts, drivers, extensions, software resources, it could be any of a number of things.

Assuming you feel strongly that a live demo is the way to go, how do you mitigate these risks?

A demo of a personal device, like a new phone or tablet usually works best if the audience can see it up close. Do realize that "up close" doesn't always have to mean within touching distance of the audience.

> *In January of 1997, Steve Jobs had to demonstrate the very first iPhone (with a 3.5-inch diagonal screen—tiny by today's standards) to an audience of 5,000 people who had never seen it before. Apple had to devise a way of showing the phone's groundbreaking interface on a screen 20-feet high for everyone to see.*

The rest, as you might know, is history. But unless you have the same resources Apple has, you might want to try a much simpler, foolproof approach.

If you're doing a live demo on a new application still undergoing testing, the last thing you want is for the app to embarrassingly crash in the middle of your presentation. And unless your app is way past the testing stages, I'd also be wary of putting it in the hands of your audience to play with.

You can avoid many of the risks by doing a show-and-tell rather than a live demo. I suggest taking screenshots and using slides to simulate your demo. If you have master-level PowerPoint skills, you could even create hot spots on the screen that seem to respond to hovers and clicks to jump to other slides, making it look more interactive and closer to the actual experience.

With a show-and-tell version of your live demo, you don't have to worry about the unexpected. No system crashes, lost connections and embarrassing missteps. All the people in the audience can see the demo up-close without squinting. You remain in full control of the outcome of a live demo.

Done correctly and in the right setting, a live demo adds yet another layer of credibility to your presentation. It can send a clear and punchy message that can erase any lingering doubts.

BUT, live demos can be risky. Be warned.

My recommendation? Do a show-and-tell instead.

CHAPTER 18
Speak their language

This chapter goes beyond the standard definition of a language. It goes beyond just avoiding jargon, speaking plainly and using analogies. "Speaking their language" means understanding and addressing what's important to your audience—using words that will resonate with them.

The audience may listen and understand what you're saying but still not agree with your conclusion. You may be presenting a set of information and arguments that you believe are important and sufficient, but your audience is looking for an entirely different set.

For example, it's not unusual for a board to expect rigor or completed staff work. They expect you to have done your research, drawn up and analyzed options, run your numbers, talked with key stakeholders, considered risks, and have a defensible recommendation. If you're asking them for a decision, they expect to have all the pertinent information they need to make that decision. This, too, is part of their "language."

As another example, let's revisit the IT executive in Chapter 2. He is presenting his IT strategy for the next 5 years to his board of directors, along with a corresponding capital budget. He ends his presentation with a request for approval.

He's done everything we've discussed so far—easy on the eyes, easy on the mind, engaging in a conversation, and so on. There's now a good chance the board understood his presentation. Yet, he still didn't get the all-important approval.

Why not?

When designing a presentation, you have to ask yourself "What is important to this particular group? How do they process information? What would make them comfortable with my proposal and be confident that approving my request is the right decision?"

For one, you must speak their language. This means thinking like them and using words that resonate with them.

Start by explaining how your IT strategy fits in the overall mission of the company. You position your budget as an investment, with returns on customer service and staff productivity. Talk about the risks if it's not implemented. Demonstrate the options you've considered and studied. Provide hard, defensible numbers where you can. Cover all the bases then make a logical recommendation ready for a decision.

Strategy. Mission. Investment. Customer service. Staff productivity. Analysis of options. Risks. Numbers. NOW you're speaking their language.

You've addressed the likely issues (and maybe some they haven't even considered) in their minds using words they were expecting. Because you've placed yourself in their shoes, they are more likely to see the value in your presentation. Now they can be comfortable giving their approval.

To know how to think like your target audience and what's important to them, you may have to do some homework.

The country manager (CM) of a multinational telecom company was invited to provide a technology brief for the executives of a large government agency. A technology brief typically benefits the executives by updating their awareness of the latest technologies that could be useful to the organization. This face-to-face also benefits the company by raising awareness of its products and solutions to influencers and decision makers.

As I had once worked with the agency, the CM asked me for help and advice.

Other competing technology companies were also doing technology briefings. I've sat through enough of them to know how it usually goes. The challenge was to stand out, make a positive impression, and plant the seeds for a business relationship.

I said, "You should set yourself apart from the usual technology briefings. These are often outright sales pitches to potential clients. Hard selling will turn off this audience. You need to be different by making yourself more relevant."

I added, "Do some homework before the technology briefing. Know what the pain points in the agency are. That way, you can talk about technology and solutions that address those pain points. You're creating

relevance by recognizing an existing concern and matching it with an appropriate solution. You will not only be remembered, you will also gain more goodwill from the agency."

" You'll never know. After learning of your technology, they just might create a project that solves a current and real problem leveraging that very technology."

I arranged for the CM to meet and interview a number of key executives one-on-one. The CM and I discussed his consolidated findings and insights. We brainstormed what to present, how to present it, then proceeded to develop his slides.

The briefing went very well. After the presentation, one executive approached the CM.

" You know, we've had many technology briefings these past years. You are the first vendor to present technologies and solutions that seem to know and address what we need."

The CM smiles. Of course it did. We designed the presentation to do just that. He did his homework. He could then "speak the language" of the executives. The CM didn't say "these are your problems, here are my company's solutions." He didn't have to. He just highlighted solutions that he already knew were relevant to the agency. On their own, the executives recognized the corporate pain points that his company's technology solutions could address.

The CM was selling, true, but in a much more subtle and effective way.

And yes, that presentation did lead to a project that addressed a pain point.

Do some homework. Know where the pain points are. Make yourself relevant.

CHAPTER 19

Bad practices

There are three common practices that actually work against the presenter. People who've heard me explain why are often surprised considering how widely accepted these practices have become. Open your mind and hear me out.

Advanced handouts

Audiences like having handouts. Providing printed copies of your slides as handouts allows the audience to better follow your presentation. They don't have to take notes. They can use the handout slides to review your talk at a later date. It gives the impression that you and the organizers are well-prepared.

Or so the usual justifications go.

My advice? Lose the handouts.

In a nutshell, giving out handouts ahead of time, even right before you start, is the worst possible form of self-inflicted sabotage you can imagine. Yes, you read that right. It provides all sorts of distractions to your audience. It gives them the perfect excuse to not establish eye contact—

which we've already determined to be absolutely essential for engagement. And it completely obliterates any element of surprise you might have carefully crafted and prepared.

A slide, if prepared properly, should be devoid of complete sentences and explanations. It should be designed as an adjunct, a backdrop, an accompaniment to a live narration. It's a fragment of a much bigger thought that can only be completed by a presenter's narrative. In other words, in the hands of person who is not in the presence of the speaker, it would be almost meaningless.

Given a copy of your slides in advance, your audience will tend to read ahead and form impressions and conclusions without the benefit of the context and explanation you have yet to provide. And when they're too busy reading, guess what? They'll be too busy to look and listen to you as well.

After a quick scan, some might decide it's not interesting enough and leave the room. They might think they've already got the message and not pay too much attention. They could also use this "spare time" to start poking holes in your arguments and wait in ambush.

What's worse, these risks can be exacerbated in three particular situations: If the organization has unhealthy corporate politics with informal groups aggressively defending turfs and interests; if you are proposing something controversial that entails significant change; or if you are proposing something that's never been done before or has failed in the past. Threatened turfs and interests, controversy, change and fear of the unknown, all tend to generate resistance.

Generating that resistance ahead of time is like committing presentational suicide.

Moreover, advanced copies will erode your control of the messaging and your audience.

Senior management, in particular, likes having copies way in advance. Why do you think that is?

The implied reason here is so they can come prepared. If they like what they see, great. If the proposal requires a decision at the meeting, they might have their staff prepare a study or, at the very least, a list of concerns and questions. They don't want to be caught flat-footed should you suddenly make a claim that they couldn't confirm on the spot. This might be reasonable, to a degree, but executives who really know their stuff won't need it.

The risk is if, based solely on your slides, they don't like what they see. They could dig in and take an opposing position without the benefit of an explanation from you.

They can even prepare to shoot you down before you even get out the gate. Any way you cut it, advanced copies of your presentation simply aren't a good idea.

But what if senior management insists? Obviously, you have to comply.

But there is still a way you can mitigate the risk of getting shot down in advance. I suggest preparing a separate, accompanying document—a synopsis which carefully paraphrases each section in proper sequence, listing rationales, examples, numbers, justifications, and what not. Keep it short and ultimately readable—one or two pages at most. Your hope is that they will read this document first and not miss out or misinterpret anything, as would be the case if they went through your slides by themselves.

The synopsis also serves another purpose. A more important one, actually. It's the handout you *should* be giving out at the end of every presentation instead of multiple photocopies of your deck. It's ideal for reviewing because it's short; it covers all the important points of the presentation; and most importantly, it doesn't have information gaps that a slide deck would.

Let's face it. Do you honestly think these executives have the time to leaf through a 60-plus-page printout of your deck just to review what you said? I'm sure they have better things to do with their time. My bets are, it either goes to a forgotten pile in the office, or worse, straight into the recycling bin. A one to two-page synopsis, on the other hand, is more likely to get read.

Besides, it's also more environmentally friendly. Think about it. If there were 10 people present, and your deck is 60 pages long, that's 600 reasons to call your company environmentally unfriendly. Ten sheets? Not so much.

Audience questions in mid-presentation

My next recommendation—no audience questions during the presentation.

You might think, what? First you suggest to not give advanced copies of the presentation. Now, you don't encourage audience questions during the presentation? That's right. Questions during the presentation, no. Questions after the presentation—absolutely.
I'm always surprised when a presenter says, "Please feel free to interrupt and ask questions at any point in the presentation."

If you're in an informal environment with an open-ended time allotment, surrounded by a familiar, friendly audience, then fine. That could work. If you're the boss, that could work too.

But as standard presentations go, time is usually limited. How are you expected to stick to your plan if the audience keeps interrupting? How many interruptions can your talk time accommodate before things go off the rails?

A better strategy would be say at the start, "I'd like to go through my entire presentation first. Please reserve your questions for the end. We'll have time for some Q&A after."

Sometimes, especially when presenting to management, someone will interrupt you anyway. You could say "I'll be getting to that in the following sections" if indeed, that's the case. Or you might give a quick answer, nothing elaborate, then move on.

Not dealing with an interruption expeditiously could result in lost time, losing your train of thought or dropping audience engagement. I've seen presentations go sideways because the presenter was unable to control the ensuing discussion that a question triggered. Interruptions of this type are disruptive and can throw the entire presentation off-topic, so be on your guard.

Clearly stating that there will be time for questions after the presentation has three advantages: It shows a willingness to engage the audience; questions in the audience's mind at one point may have already been answered by the end of your talk; Q&As don't eat into your talk time and can be cut short anytime.

Setting limits to the number of slides

Here's another bad practice that's becoming increasingly common—limiting presentations to a fixed number of slides as a means to control presentation time.

Earlier, in Chapter 8, I recommended that "if a slide has two messages, split it into two slides. Slides are free." That means you end up with more slides, which begs the question, how many slides should a presentation have? Is there a rule of thumb?

The correct answer really is, it depends—on the purpose, the message, the audience and your talk time. The number of slides shouldn't matter, as long as the message is delivered effectively and achieves the intended purpose within the allotted time.

However, it's become a common practice, particularly among tradeshow organizers, to set a limit to the number of slides per presenter, especially if it's a set program with a fixed number of speakers. These organizers work under the misguided notion that limiting the number of a presenter's slides will automatically reduce their talking time. The same goes for a lot of companies that claim to have hit upon a 'magic number' for the number of talk minutes per slide (which painfully ignores the different ways that people can or prefer to present information).

Don't they realize that, if really necessary, any presentation can use animation to pack any number of slides into whatever limit was set?

An IT manager presented the specifications of a School Enrollment System that they intended to develop and deploy in a network of schools. The presentation took about 20 minutes and used what felt like maybe 15-20 slides.

I asked for a copy of the PowerPoint presentation. I received a file with one slide.

That's right. The entire presentation was in just one slide!

It was a mess of a slide—it was useless trying to print it. The IT Manager used animation to cause text bullets and images to come in and out of the presentation. To his credit and lucky for him, he pulled it off. And I could click through his entire presentation. It works.

Still, there was no logical reason to put everything in one slide. It was an inefficient (and very difficult) way of doing it. The slide was also useless as printed material. I can only guess that the IT Manager was mischievously challenged to prove that he can do it in just one slide.

But it can be done.

Most slides can be presented anywhere from maybe five seconds to a couple of minutes. In the above case, 20 minutes. If I embed a video, then obviously, sky's the limit.

So really, the number of slides does not translate well to presentation time.

Rules on the number of slides and number of talk minutes per slide are unfortunate and don't make any sense.

I would say, just use as many slides as you need but not more. I would also say, use as few slides as you can get away with but not less. The number of slides should never have precedence over the purpose, the message and the time allotted to a presenter.

I would even add, if you can deliver the message effectively with zero slides, that works too.

Unfortunately, not all presenters have the skill to get things done within the time allotted. They either pack way too many slides and don't practice, or allow their minds to wander, digress and 'overtalk' some points, only to discover too late that they had run out of time and rush through the all-important close.

These folks should take to heart Chapter 3 (The elevator pitch) and Chapter 13 (Time your presentation).

So what options do organizers and companies have other than setting a limit to the number of slides?

Instead of saying "Deliver your message in 5 slides or less"—which assumes it takes a minute on average to talk to a slide—why not just set a time limit. "Deliver your message in 5 minutes or less."

Inform the presenter in advance of his allotted time. Be clear why there's a time limit. Be explicit that it's non-negotiable, and that there could be consequences, like abruptly getting the hook. Setting a non-negotiable time limit will force presenters to simplify their messages and practice their delivery. Let the unprepared learn their lesson the hard way.

A skilled presenter will achieve his purpose by delivering his intended message in the allotted time. The number of slides is irrelevant to the equation.

CHAPTER 20
Allies in the audience

When you're presenting a proposal to a board or management team, it helps to have an ally in the audience.

Your ally can vote in your favor, ask questions that can help you elaborate a weak point, make comments that support your case and initiate a motion to approve your proposal if other members hesitate.

An ally is someone whom you feel would be supportive of your proposal. You would have asked for a one-on-one discussion with this person prior to the board meeting. He or she would have seen your presentation, raised any concern and got clarification. You may have asked and been given advice on what the board might want to hear. Lastly, you would have requested that person to be present in the audience for support during the presentation.

At one time, I was engaged as a technical consultant by a government education agency. One of my roles was helping integrate various quality education initiatives into a single quality framework. The agency's five-member board was responsible for approving any proposal before implementation.

I joined the Technical Working Group (TWG) tasked to create this new framework. The TWG members were from government and the academe.

When the TWG recommendations were ready, I met with a board director who was passionate about quality education.

I presented the proposed framework. The director happily agreed to support it. She went one step further. She called in another director, a potential critic, into our meeting. After some discussion, this other director also agreed to support the proposal.

A few days later, I presented the proposal to the board.

After I finished presenting, the chairman, who sponsored the initiative, started to ask questions. He probably wanted it vetted by the board anyway.

I answered the first few questions.

Then something surprising happened. Before I could answer the next question myself, one of the directors whom I met with, decided to chime in. Later, both she and the other director were providing the chairman with answers.

I watched with amusement. I realized I had forgotten to tell the chairman that I already had the support of the two directors.

The chairman realized what was happening. He asked the two directors, "Do both of you know about this proposal?"

They nodded.

" And you support it?"

" Yes, we do."

A vote was called and the proposed quality framework was approved unanimously.

In a five-man board, one was the sponsor and two were allies. Approval by majority vote was a foregone conclusion but getting a unanimous vote was even better.

Sometimes, your ally can help pave the way for an approval even before a presentation starts.

I was assigned by the president of a government agency to take over management of a $16M IT procurement project with aggressive timelines.

I was then called to a meeting by a board director who headed the board's IT Committee. The director starts with "I understand you're now the project manager. The board wants this project bid out and awarded in seven months. Can you do it?"

I explained why that would be quite a challenge.

" In this agency, procurement of this size usually takes at least eighteen months. Three to four of those months

are spent on back-and-forth discussions with and within the 9-member board. The board will not hand down a decision until it is satisfied enough to make an award. That's why the seven-month deadline is a virtual impossibility."

I then made a bold suggestion on how we could make it happen anyway.

" I have an idea. My project team will deliver a project recommendation to the board ready for awarding in five-and-a-half months. All throughout, I will keep you updated. You then have two weeks to get the award recommendation approved by the board. We'll have one month left to spare, just in case."

The director agreed. "Deal."

The president was informed of the deal and added his support.

The project had a very successful competitive bidding. Several international vendors submitted complying bids. Happily, the vendor with the best technical specifications also submitted the lowest bid.

The project team delivered the project in five-and-a-half months as promised. I provided all the bid documents and the team's award recommendation to the board.

Two weeks later, I made a short ten-minute presentation to the board.

My director ally moved for approval. Another director seconded. Nobody objected. The award was approved.

No questions. No discussion. Just a quick and easy approval.

How the team managed to do in six months what would normally have taken eighteen, is a project management case study in itself.

But it was the swift approval by the board that surprised everybody.

Apparently, my director ally was communicating the project's progress (and our deal) to the board. I assume he also had the board's legal counsel review the recommendation documents. The board was aware and satisfied with the conduct and results of the procurement even before I presented.

My ten-minute presentation was a formality.

CHAPTER 21
Closing act

You've gone through all your slides and messages. It's time to conclude your presentation. You have to do two things—summarize your key points and, in the case of a proposal, get the win.

Summarize

When summarizing your key points, be clear and concise. There are two reasons for doing a summary.

First, you need to remind the audience of your core message—the takeaway from your presentation. This is what you want them to remember after they leave.

Second, it sets the stage for you to make your request and get your win.

Get the win

Remember how, in your opening, you told the audience what the purpose of your talk was and what you will need from the audience when you finish?

Now is the time to actually make your request. What do you specifically want them to do now? Consider your recommendation? Approve your proposal? Change the status quo? After all, that is the whole reason for your talk.

Over lunch, a colleague was telling me about a presentation he was preparing to deliver before his unit's managers.

I then asked, "At the end of your presentation, what would you like to happen?"

" The director and his managers would be aware of the various initiatives my group is doing, and that we're doing a good job."

" Just that? You just want them to know about your initiatives?"

" Well, no. If all goes well, maybe I'll have a chance to ask them to give my group additional resources. But I can do that in a future presentation."

" Why would you want to do that in a future presentation? If that is your win, why don't you ask for it now?"

He pauses, reflects. "You don't think that's too pushy?"

" No. Maybe they'll approve your request. Or maybe they'll tell you what you need to do first so that they'll approve it the next time you ask. Worst case, they'll say no. Either way, you get some clarity for moving forward by asking."

I then offered my usual advice. "You should always end your presentation with your win. If getting additional resources for your group is your win, you should close your presentation by asking for it."

I'm always surprised when a presenter doesn't plan on closing his presentation with a win.

After you've made your summary, you might say "When I began this talk, I said that I would need your commitment to…" or perhaps, "I am now asking for your approval…" It's surprising how some people will spend all the effort preparing and delivering a presentation, only to forget (or chicken out from) stating the actual request, explicitly.

The only reason I would forgo the request is if I felt that I've failed in persuading my audience to arrive at a decision. Perhaps there were too many objections raised or too many questions that I couldn't answer. Asking for approval in such a situation would be futile.

If that were the case, I would deviate from script and instead, thank the audience for their inputs, and commit to use them to improve the proposal.

Here's a story that ends with a strong close, but also incorporates some of the earlier lessons in this book.

> *Circa 2000, a school network offered distance learning (DL) programs in computer programming and electronics technology. Back then, distance learning meant mailing printed study materials to students, with validation exams conducted at the school's branches.*

The academic head for the network, a friend, calls me up. "I hear the president is going to ask you to study why our distance learning program is unprofitable. You just might get assigned to fix it. I suggest you don't take that on."

" Why not?", I asked.

" Several managers more senior than you have tried and failed to make that program work. Regardless what you do, they will all have something to say. And I don't want you to be the next to fail."

True enough, at the next management committee (ManCom) meeting, the president directed me to study the problem. Against friendly advice, I took on the challenge.

Past DL managers agreed on the symptom of the problem—very high dropout rates—but placed the blame on the quality of the study materials. I soon realized the real reason for the high dropout rates.

Two weeks later, I presented before the ManCom. Towards the end of my presentation, I started my closing with a summary.

" Past efforts focused on improving the study materials. The materials are not top notch but they aren't bad. I'd give the materials a B+. It is a concern but that, however, is not the root cause of our problem."

Acknowledging that the materials weren't perfect gave past DL managers some validation that they weren't all wrong. I then continued with a key insight that they missed, but were now curious and open to listen to. There was no reason to shoot me down … yet.

" *We've designed the program to be very easy for anyone to get started. Initial fees are very low so that we attract many new students. However, low initial fees also meant they weren't invested enough. Plus, we've done NOTHING to keep the students motivated to stay with the program and pay the subsequent fees."*

Now speaking slowly, I delivered the key insight.

" *It's just as easy for the unmotivated to drop out as it was for them to get in. The result? Our dropout rates are just too high for the program to be profitable."*

I paused for a few seconds to allow that insight to sink in, and scanned the room. All eyes were on me, waiting for my next word. I could see that nobody was moving to react. Before anybody could raise a point, I continued.

Speaking deliberately, confidently, emphatically, and with eyes locked with the audience, I delivered the final closing spiel that I've repeatedly practiced and visualized.

" *UNLESS AND UNTIL we shift our efforts from improving teaching materials to motivating our students to stay with the program, dropout rates will remain high. This shift however, will impact the way we run our schools and retain our students."*

Short pause. Now for the recommendation.

" *IF WE ARE UNWILLING to adopt this change in all our schools, I recommend we close down the distance learning program."*

Silence. They were all expecting a fix. Nobody expected a conditional recommendation to close down the program. Even the president said nothing. He then adjourned the meeting.

One of the managers who previously handled the program came up to me. "That was an insightful analysis. You're right."

Another nodded in agreement, adding "and that was a very strong close."

I could still hear myself emphatically saying, "UNLESS AND UNTIL…"

Soon after, the school network closed down its DL program.

Epilogue

At the end of a presentation, how do you gauge if it went well? Are there certain audience reactions to look out for that suggest you've succeeded?

Are you looking for people to raise their hands and ask questions or would you rather have them turn quiet and not have any? The answer really depends on what the purpose of your presentation is.

If you're selling an idea and want to get the audience interested and excited about it, then raised hands with questions would be ideal. You can usually measure engagement from the type of questions asked.

If I'm presenting something with the goal of making the audience understand, say, the status of a project or the result of a study, what response should I be looking for? Most presenters would probably welcome a few questions, taking these as signs of interest and engagement. By the same token, if the questions revealed key points that I missed, I'd be somewhat disappointed in myself for the oversight.

But sometimes, everything just falls into place. From the preparation, to the delivery, to the request for approval, that in the end, the audience is quiet, attentive, but with no questions.

When that happens, I break the silence with "Are there any questions? No questions can only mean you either understood nothing or you understood everything." In most cases, my audience would laugh and assure me that everything was clear and that they have no questions.

Lastly, if I'm presenting a proposal, with the end purpose of asking for the audience's explicit approval of the proposal, what is the ideal response?

Time for one last story.

The room turned quiet. The directors on the board are all silent. All eyes are on you. You finished presenting a budget proposal for a high-profile technical project. You had just asked the board to approve a $60 million budget.

The government project had attracted media attention and the board was anxious. The proposed budget must pass scrutiny in any forum and this was the first time a budget number has been proposed.

The chairman had a faint smile and looked pleased with the silence. He had grown to expect a barrage of pointed questions after each presentation, even for proposals with a much smaller request and less potential for controversy.

This presentation was different. It had clarity, bulletproof logic and unimpeachable rigor.

He could see that his directors were all in deep thought, searching for questions... probing for a weakness ... to clarify a point.

Nothing.

A few more seconds, one director breaks the awkward silence and moves for approval. Three seconds of silence followed. Another director seconded. No objections.

The president flashes a big smile. You can't help but smile yourself.

Slam dunk.

Sources

FOREWORD

Pink, Daniel H. To Sell Is Human: The Surprising Truth About Moving Others. New York: Riverhead Books, The Penguin Group, 2012.

CHAPTER 2

Netmarketshare. "Market Share Statistics for Internet Technologies." Last modified, November, 2017. http://www.netmarketshare.com

CHAPTER 5

Lee, Dick and Hatesohl, Delmar. University of Missouri Extension. "Listening: Our Most Used Communication Skill." Reviewed, October 1992. http://extension.missouri.edu/p/CM150

Rayner, Keith and Clifton, Charles Jr. "Language Processing in Reading and Speech Perception is Fast and Incremental: Implications for Event-related Potential Research." U.S. National Library of Medicine, National Institutes of Health. Published online: May 15, 2008. https://www.ncbi.nlm.nih.gov/pmc/articles/PMC2649675/

Medina, John J. Brain Rules. Seattle: Pear Press. 2014

Bowan, Merrill D. Integrating Vision with the Other Senses. Neurodevelopmental Optometry. Drafted August 8, 1999. http://www.nb.net/~sparrow/integrate.html

Miller, George A. "The Magical Number Seven, Plus or Minus Two." Psychological Review. Princeton University Department of Psychology. 1956.

Part III

Thoreau, Henry David. Walden; or, Life in the Woods. Boston: Ticknor and Fields. 1854.

CHAPTER 10

Bonesteel, Matt. "Nike lost Stephen Curry to Under Armour because it got lazy." The Washington Post. Published online: March 23, 2016. https://www.washingtonpost.com/news/early-lead/wp/2016/03/23/nike-lost-stephen-curry-to-under-armour-because-it-got-lazy/?utm_term=.d0fdf1572436

CHAPTER 13

Gallo, Carmine. The Presentation Secrets of Steve Jobs: how to be insanely great in front of any audience. USA: McGraw-Hill, 2010.

CHAPTER 17

"Bill Gates releases mosquitoes into audience.", NBC News.com: Infectious Diseases. Updated February 4, 2008. http://www.nbcnews.com/id/29022220/ns/health-infectious_diseases/t/bill-gates-releases-mosquitoes-audience/#.WiiKmrYZOJE